AF785490

Advances in Modelling and Control of Wind and Hydrogenerators

Edited by Amir Ebrahimi

Published in London, United Kingdom

IntechOpen

Supporting open minds since 2005

Advances in Modelling and Control of Wind and Hydrogenerators
http://dx.doi.org/10.5772/intechopen.77988
Edited by Amir Ebrahimi

Contributors
Asma Aziz, Ievgen Zaitsev, Anatolii Levytskyi, Marwa Hassan, Shady H.E. Abdel Aleem, Foad H Gandoman, Abdollah Ahmadi, Masoud Ardeshir, Ali Esmaeel Nezhad, Joeri Van Mierlo, Maitane Berecibar, Ayman Attya, Jose Luis Dominguez-Garcia, Adelhard Beni Rehiara, Naoto Yorino, Yutaka Sasaki, Yoshifumi Zoka, Faisal Wani, Jianning Dong, Henk Polinder, Babette Schwarz, Bastian Diebel, Axel Walter-Krause, Roland Jester-Zürker

© The Editor(s) and the Author(s) 2020
The rights of the editor(s) and the author(s) have been asserted in accordance with the Copyright, Designs and Patents Act 1988. All rights to the book as a whole are reserved by INTECHOPEN LIMITED. The book as a whole (compilation) cannot be reproduced, distributed or used for commercial or non-commercial purposes without INTECHOPEN LIMITED's written permission. Enquiries concerning the use of the book should be directed to INTECHOPEN LIMITED rights and permissions department (permissions@intechopen.com).
Violations are liable to prosecution under the governing Copyright Law.

(cc) BY-NC

Individual chapters of this publication are distributed under the terms of the Creative Commons Attribution - NonCommercial 4.0 International which permits use, distribution and reproduction of the individual chapters for non-commercial purposes, provided the original author(s) and source publication are appropriately acknowledged. More details and guidelines concerning content reuse and adaptation can be found at http://www.intechopen.com/copyright-policy.html.

Notice
Statements and opinions expressed in the chapters are these of the individual contributors and not necessarily those of the editors or publisher. No responsibility is accepted for the accuracy of information contained in the published chapters. The publisher assumes no responsibility for any damage or injury to persons or property arising out of the use of any materials, instructions, methods or ideas contained in the book.

First published in London, United Kingdom, 2020 by IntechOpen
IntechOpen is the global imprint of INTECHOPEN LIMITED, registered in England and Wales, registration number: 11086078, 7th floor, 10 Lower Thames Street, London, EC3R 6AF, United Kingdom
Printed in Croatia

British Library Cataloguing-in-Publication Data
A catalogue record for this book is available from the British Library

Additional hard and PDF copies can be obtained from orders@intechopen.com

Advances in Modelling and Control of Wind and Hydrogenerators
Edited by Amir Ebrahimi
p. cm.
Print ISBN 978-1-83880-532-6
Online ISBN 978-1-83880-533-3
eBook (PDF) ISBN 978-1-83880-544-9

An electronic version of this book is freely available, thanks to the support of libraries working with Knowledge Unlatched. KU is a collaborative initiative designed to make high quality books Open Access for the public good. More information about the initiative and links to the Open Access version can be found at www.knowledgeunlatched.org

For EU product safety concerns:
IN TECH d.o.o., Prolaz Marije Krucifikse Kozulić 3, 51000 Rijeka, Croatia, info@intechopen.com or visit our website at intechopen.com.

We are IntechOpen, the world's leading publisher of Open Access books

Built by scientists, for scientists

4,700+
Open access books available

121,000+
International authors and editors

135M+
Downloads

151
Countries delivered to

Our authors are among the
Top 1%
most cited scientists

12.2%
Contributors from top 500 universities

WEB OF SCIENCE™

Selection of our books indexed in the Book Citation Index
in Web of Science™ Core Collection (BKCI)

Interested in publishing with us?
Contact book.department@intechopen.com

Numbers displayed above are based on latest data collected.
For more information visit www.intechopen.com

Meet the editor

Amir Ebrahimi received his PhD in electrical engineering from the University of Stuttgart, Germany with a doctoral dissertation on analytical modelling and optimization of surface mounted permanent magnetic synchronous motors considering spatial harmonics. From 2013 to 2017, he was group manager for electrical machines at the Fraunhofer Institute for Manufacturing Engineering and Automation in Stuttgart. Since November 2017, he has been a professor for electrical machines at the Institute for Drive Systems and Power Electronics at Leibniz University Hannover. His research interests are analytical and numerical calculation of transient process in large electrical machines, particularly hydro and wind generators. Prof. Ebrahimi is a reviewer of different IEEE transactions and a member of the European Energy Research Alliance.

Contents

Preface

The global demand for power has grown by more than 50% in the last 20 years and it will increase even more drastically in the next few years with an expanding trend for electrifying our lifestyle, for example e-mobility and diverse electronic devices. Apparently, mother nature cannot continue providing us with her generous hospitality if we do not promptly rethink our energy generation concepts and take all measures necessary to switch to clean and renewable energy with no or at least limited environmental impact. This is no more a fantasy but a crucial first priority demand, which, if not thoroughly treated now, will leave irreversible consequences and this is my concern. However, the sole subsidizing of wind and solar energy is not yet a sustainable solution. It could be the solution if we thoroughly prepare our electrical infrastructure to deal with the renewable energy penetration.

Serious renewable energy systems mostly focus on wind energy and it is easily forgotten that water (hydropower) is not only the main source of renewable energy but also the most efficient one. Hydropower is an incontestable renewable energy source, especially since it meets government agendas for sustainable greener energy sources providing substantial savings in CO2. Finally, when it comes to cost and reliability, nothing compares to the cost of electricity coming from a hydropower plant. Hydropower plants are the most important key factor of a sustainable electrical grid.

Rapid deployment of wind and solar energy generation is going to result in a series of new problems with regards to the reliability of our electrical grid in terms of outages, cost, and life-time, forcing us to promptly deal with the challenging restructuring of our energy systems. Increased penetration of fluctuating renewable energy resources is a challenge for the electrical grid and there is a need for a backup capacity and energy storage. Pump Storage Power Plants (PSPP) are the most efficient and largest commercial energy storage systems. Their main function is to support the electrical grid during fluctuations that could be mostly caused by wind energy penetration in the near future. Developing additional hydropower pumped storage, particularly in areas with recently increased wind and solar capacity, would significantly improve grid reliability. PSPP plants have to be available on demand to quickly balance load fluctuations, i.e. inject fast power to the grid in peak load times or extract power from the grid in times of overcapacity, e.g. when high amounts of solar or wind power feed the grid. The best thing about this solution is that one renewable energy source supports the other one. However, there are at least two decisive challenges to be overcome. First, most of the PSPPs are provided with fixed-speed generators and cannot absorb or inject an arbitrary amount of electrical power into the grid. Second, traditional PSPPs are not agile enough to compensate the dynamic fluctuation of modern electrical grids. Therefore, one of many advances during the last decades has been the development of adjustable speed systems to allow for controllable power in the pumping and generating mode. Solutions are either to use other types of machines, e.g. Doubly Fed Induction Generators (DFIG) instead of the Synchronous Generators (SG) or to provide the generators with Power Electronic Converters. However, both solutions save impacts on the functionality, life-time, and reliability of the generator due to interaction influences. The deployment of wind and solar energy is primarily determined by our capability to comprehensively handle these new interactions, which results in more reliable generators and consequently increases the acceptance of renewable energy. We must find methods to deal with aspects such as permanent start-stop operation, power electronics interactions, monitoring, life-time estimation, and failure prediction, if we want to move towards reliable renewable systems. High-quality maintenance and monitoring are essential to ensure a high level of availability of large generators, since unexpected outages result in huge economic losses, limit the availability, and impair the reliability of these systems. The latest monitoring is a set of offline, online, and intelligent methods based on measurements and expert knowledge. The main deficit of these methods is the lack of comprehensive multi-physical modelling of different phenomena. Since there are no methods to deal with these multi-physical phenomena, we have focused on developing methods for indirect anticipation of operating conditions. Consequently, these traditional methods are only applicable to a certain insulation system or test condition, which is still a subject of debate among researchers. The latest modified monitoring method is to implement accelerating aging tests, characterize the parameters, and verify the model in the lab. It is worth mentioning that even on the lab scale, the test objects are exposed to controlled thermal, electrical, mechanical, and thermal cycling stresses almost independently, which is not the case in the real generator.

This book covers new technologies and modelling methods of renewable energy generators including wind, ocean, and hydropower systems in two sections. In the first section, three topics related to hydro energy systems are presented. The second section deals with wind energy systems and issues related to the volatile nature of renewable energies. The control systems of wind power systems for better frequency control and grid stabilization are included in this chapter.

The first chapter is dedicated to computational fluid dynamic simulation of large hydro generators. The CFD model to be validated consists of the full generator geometry, which is modeled in high geometrical detail. The steady state multiple reference frame approach was chosen for the simulations and the influence of different rotor-stator interfaces and turbulence models was investigated.

Chapter two discusses the hybrid electro-optic sensors for the fault diagnostic system of hydrogenerators. The application of hybrid electro-optic sensors (HFOS) with capacitive mechanical sensors in hydrogenerators for fault diagnosis, measurements, and parameter extractions are described in this chapter.

Besides the traditional hydropower plant, there are also non-conventional methods to use the hydro energy from the oceans.

Chapter three gives an overview of the functionality of tidal generators and compares the flooded and sealed generators.

Chapter four gives an overview of onshore and offshore wind energy technologies. Different components of wind farm as well as the technologies used in them are investigated and possible layouts regarding the foundation of an offshore wind turbine, floating offshore, as well as the operation of wind farms in the shallow and deep location of the ocean are studied.

Chapter five presents an adaptive load frequency control based on the least square method. The controller adopts an internal model control (IMC) structure in two scenarios, i.e. static controller gain with adaptive internal model and both the adaptive controller gain and adaptive internal model. A two-area power system is used to test and validate both performance and the effectiveness of this controller through some case studies.

Chapter six describes the supplementary controls to provide ancillary services. It exploits key examples of these controllers and considers their integration into the conventional control of renewable generators.

A frequency processor-based frequency-active power set point control architecture for variable speed wind turbine generator is presented in Chapter seven. Grid frequency processor based on moving averaged frequency and dynamic dead-band is tested for two different grid codes. Generated active power set point is provided to a modified torque-pitch control loop in Type III and Type IV variable speed wind turbine generator generic models.

Chapter eight investigates the effectiveness of the non-linear control-based model and the sampled-data design through the power system application. In particular, the study focuses on a model of a wind turbine system fed by a Doubly Fed Induction Generator.

Amir Ebrahimi
Professor,
Leibniz University Hannover,
Germany

Section 1

Hydrogenerators

Chapter 1

Validating a CFD Simulation Approach by Ventilation Measurements for an Air-Cooled Salient Pole Model Generator

Bastian Diebel, Axel Walter-Krause,
Roland Jester-Zuerker and Babette Schwarz

Abstract

Pressure and flowrate measurements were conducted for an air-cooled salient pole hydro model generator in order to validate a computational fluid dynamics (CFD) simulation approach. The ventilation system of the model generator was driven by adjustable external fans, which allows detailed pressure measurements for a range of operating conditions. The CFD model to be validated consists of full generator geometry which is modelled in high geometrical detail. The steady-state multiple reference frame approach was chosen for the simulations, and the influence of different rotor-stator interfaces and turbulence models was investigated. The comparison of measurement and simulations includes the static pressure along the flow path through the machine, the performance map of the external fans, and an analytical approach to describe the dimensionless machine parameters of the model generator. Good overall agreement was found between measurement and CFD, which justifies the application of the presented simulation approach in the design of ventilation and cooling systems for hydro power generators. Qualitatively the CFD simulations reproduced all measured flow effects. Also quantitatively a good prediction of measured values was identified for a broad range of operating conditions. However, it was found that the simulation accuracy does not only depend on the numerical models in use but also on the specific operating conditions and their affiliated airflow characteristics.

Keywords: electrical machine, salient pole hydro power model generator, cooling and ventilation, computational fluid dynamics, rotor-stator interface, turbulence modelling, flow and pressure measurement

1. Introduction

Today's generator technology is highly efficient. Nevertheless, the remaining electromagnetic losses in the range of 1% of the generator's rated power produce a considerable amount of heat. To remove this heat from the machine, a safe and reliable ventilation and cooling system is designed during the layout process individually for each hydro power generator.

During commissioning of many hydro generators, the ventilation and cooling system is evaluated as a part of the efficiency measurement. In the case of detecting

IntechOpen

insufficient cooling performance at this stage of the project, changes of the ventilation design become expensive and time-consuming. Therefore, reliable tools for developing the cooling system early in the design phase are required. Within recent years, computational fluid dynamics (CFD) simulations became a commonly used tool for such applications. However, in order to achieve the required calculation accuracy, detailed validation of the simulations is mandatory.

Normally, during commissioning or operation of hydro power generators, the opportunities for detailed measurements are limited, especially due to unavoidable standstill for installation and removal of measurement equipment. To overcome this problem, Voith Hydro operates a fully functional salient pole model generator which allows detailed measurement campaigns for different machine configurations without the limiting boundary conditions that are found in operational hydro power plants. This includes but is not limited to the investigation of the ventilation and cooling system of the model machine.

For this model machine, the airflow of the ventilation and cooling system was investigated in detail. The static pressure was measured at 17 representative locations within the machine. Additionally, the flowrate was monitored for each of the two fans that are mounted on the top of the test rig in order to drive the ventilation circuit. To obtain the correct thermodynamic state of the cooling air, temperature, barometric pressure, and relative humidity were measured. Different rotational speeds of the electric machine as well as of the fans were investigated to evaluate the ventilation performance within a broad range of operating conditions.

In addition, CFD simulations using the commercial software Star-CCM+ were performed and compared to the measurements. The simulation model included the complete geometry of the model machine, with all relevant parts modelled in high geometrical detail. This allows a direct comparison with all available measurement locations. Instead of modelling the fans, the measured volume flowrate was specified for the respective operating point. The air-to-water cooler of the test rig was modelled as porous media. The steady-state multiple reference frame approach was used for all simulations. The influences of different modelling methodologies such as the choice of rotor-stator interfaces and turbulence models were investigated.

In the following, the measured machine configuration is described in detail. Afterward, the measurement setup and the simulation model are presented. Finally, the measurement data is evaluated, interpreted, and compared to the simulations. This includes the discussion of the static pressure along the flow path through the machine and the performance map of the fans. Also the machine pressure loss is evaluated by means of dimensionless parameters. For all presented data, the focus is on the comparison between CFD and measurement results.

2. Model generator configuration

As described in the previous section, measurement data for large hydro power generators is hard to obtain, as restrictions in accessibility and modifiability of operational machines do not allow extensive measurement campaigns.

To overcome this problem, Voith Hydro has developed a small-scale model generator that might be operated similar to representative machines for large hydro applications. The main objectives of this test rig are to obtain measurement data for the validation of design tools but also to develop and test new innovative products and product improvements for hydro power generators. Although a clear focus during the design of the test rig was on electromagnetic similarity, investigations in the field of generator cooling and ventilation are possible as well.

The following subsections give a short overview of the model machine and its investigated cooling and ventilation configuration.

2.1 Overview

The investigated model generator is a synchronous, salient pole electric machine as it is typical for hydro power applications. On the test rig, the shaft of the generator is connected to an electric motor that feeds mechanical power into the system. The electric side of the generator is connected to a frequency converter-driven grid simulator, which allows numerous operating conditions at rotational speeds up to 750 rpm and a maximum electric output of 1.1 MVA.

An overview picture of the model generator is shown in **Figure 1**. The machine is completely encapsulated and thermally insulated from the environment. The air cooling circuit is driven by two parallel radial fans that are mounted on the top of the test rig. One air-to-water cooler removes the heat from the system.

2.2 Ventilation design

As it is typical for salient pole hydro generators, the investigated model generator is fully air cooled. **Figure 2** shows a schematic of the ventilation design, with blue arrows indicating the flow path.

The airflow is driven through the machine by two adjustable radial fans mounted on the top of the test rig. On the pressure side of the fans, the air enters the machine on both sides in axial direction. From there on it splits into two different

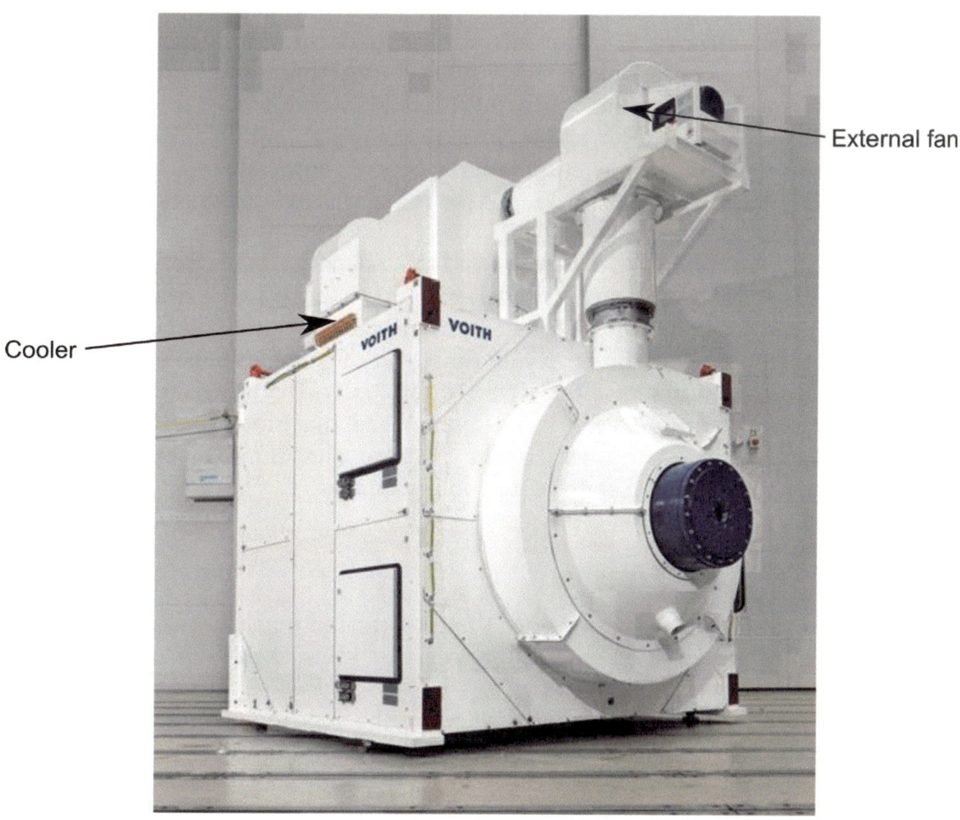

Figure 1.
Model generator overview.

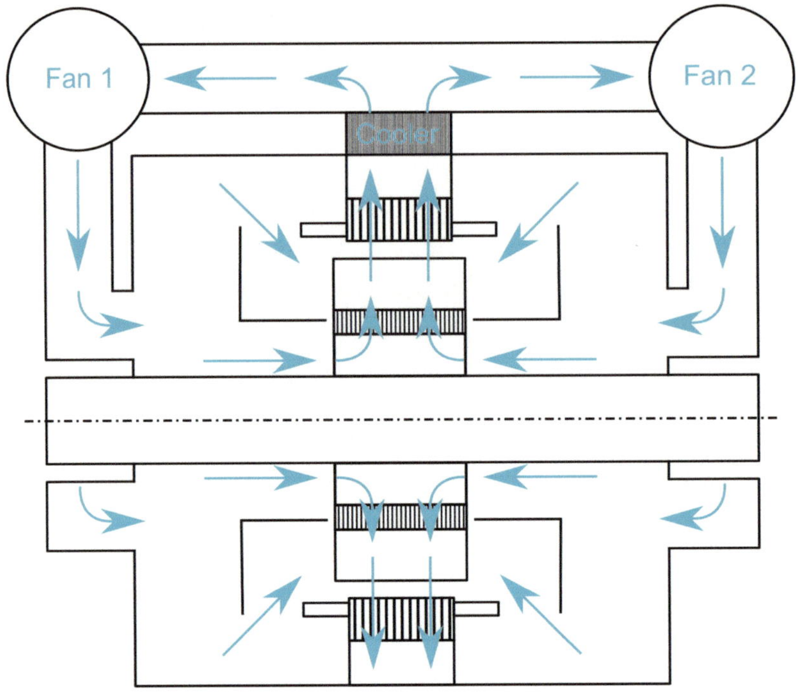

Figure 2.
Model generator ventilation scheme.

flow paths. The air on the first path directly enters the pole gaps and the annulus between rotor and stator (air gap) in axial direction. In the second path, the air enters the rotor hub openings close to the shaft in axial direction, from where it is guided radially outward into the pole gap via rectangular ducts. At this point, the two flow paths unite, and the air passes through the radial cooling ducts in the stator core toward the air-to-water cooler, where the heat is removed from the machine. After the cooler, the airflow separates into the two paths toward the fans, where the ventilation loop is closed.

Compared to the cooling scheme of most hydro generators in operation, two aspects are rather special about this model configuration: first, the split of flow paths in the rotor does not occur in most configurations as either all cooling air is guided radially through the rim or axially through the pole gaps. Second, most operating hydro generators have the air-to-water coolers distributed periodically around the stator circumference. For the given model machine, the cooler is positioned on the top of the test rig, which might lead to a less homogenous circumferential flow distribution in the stator core cooling ducts.

2.3 Operating conditions

As only the aerodynamic behavior of the investigated machine was of interest, all measurements were carried out without rotor winding excitation and in stator winding phase-to-phase short circuit.

To investigate a broad range of relevant operating conditions, measurement data was recorded for different rotational speeds of the electric machine, as well as for different rotational speeds of the two fans. Thereby, both fans were always run with the same rotational speed at a time.

Table 1 gives an overview of the investigated operating ranges. For each machine speed, measurements for four different fan speeds were done. In total,

Machine parameter		Investigated operating range
Rotational speed electric machine	$n_{Machine}$	0–750 rpm
Rotational speed fan	n_{Fan}	0–2932 rpm

Table 1.
Operating conditions overview.

measurements and CFD simulations for 12 operating points were conducted, which provides a full performance map of the model generator's ventilation system.

3. Measurement setup

The following sections shall provide a full overview of the measurement data that was obtained during the measurement campaign. Data was acquired for the static pressure at different positions of the machine, the thermodynamic state of the cooling air, and the volumetric flowrate through the model generator.

3.1 Static pressure

Static pressure was measured at 17 positions within the generator. All data was recorded relative to a common reference, which is the ambient pressure within the laboratory where the generator test rig is located. **Figure 3** indicates the pressure measurement positions within the machine. The pressure positions are named with abbreviations according to their location, which are documented in **Table 2**. The number behind the named positions indicates that multiple circumferential positions were measured, i.e., 1–2 indicates two circumferential positions, whereas 1–4 represents a circumferential distribution of four sensors.

All positions in **Figure 3** indicated with a dot represent single-spot measurements that were realized by locating an open end of a pressure tube at a position

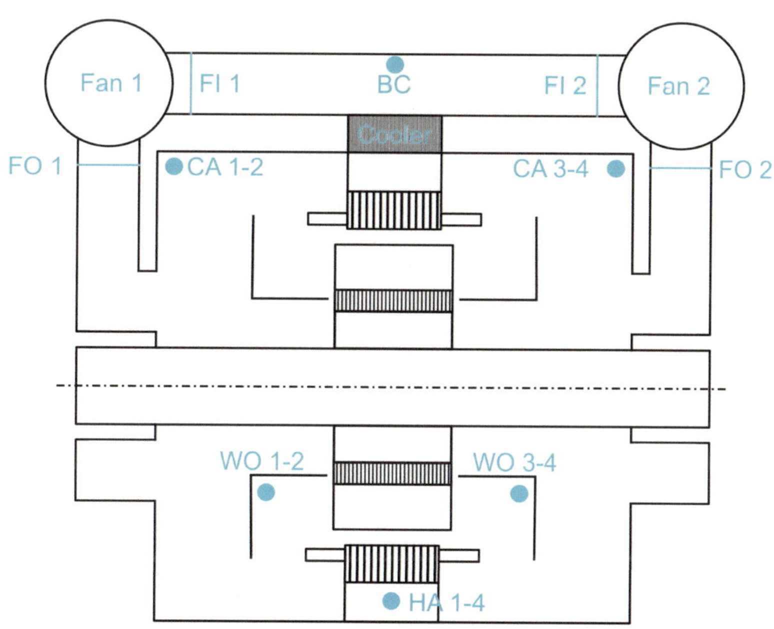

Figure 3.
Static pressure measurement positions.

where minimal flow velocities are expected (**Figure 4**). The positions indicated by a line (**Figure 3**, FI and FO) represent a wall pressure measurement at positions with higher flow velocities, i.e., before and after the fans. There, three wall pressure ports have been connected by a circular pipe to measure a circumferentially averaged wall pressure. An example for a circular pipe can be seen in **Figure 5**.

Abbreviation	Location
FI	Fan inlet
FO	Fan outlet
CA	Cold air compartment
WO	Winding overhang
HA	Hot air compartment
BC	Behind cooler

Table 2.
Abbreviations of measurement positions.

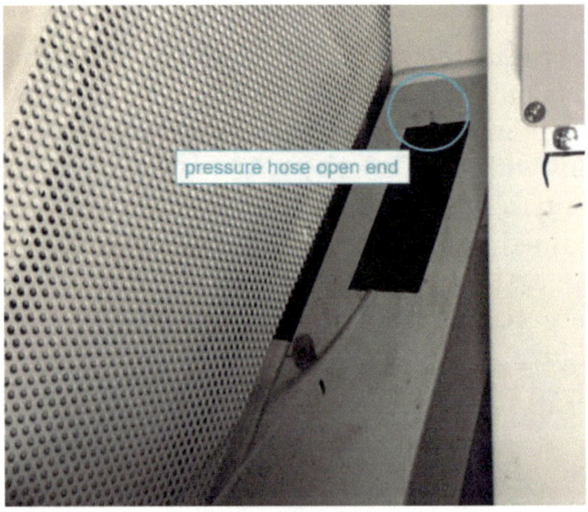

Figure 4.
Single-spot measurement.

Figure 5.
Circular pipe measurement.

All measurement data was acquired simultaneously by a multichannel pressure transducer. Data reduction was done for the recorded, time-resolved pressure data. Therefore, arithmetic averaging was done for each operating point after reaching steady-state operation. As pressure transducers tend to have a small zero drift, a correction of the recorded pressure readings was applied. To do so, the time-averaged data was subtracted by the time-averaged pressure reading at complete standstill of machine and fans. All static measurement data evaluation in this chapter is based on pressure data derived according to the procedure described above.

3.2 Thermodynamic state

The thermodynamic state of the cooling air was acquired by measuring the temperature T with a type A Pt100 temperature sensor and the relative humidity φ with a humidity sensor, both positioned on the cold air side of the machine behind the cooler. The absolute pressure p_a was measured in the laboratory. Using the data from the previously described static measurements, the absolute pressure in the closed air circuit behind the cooler was calculated using the formula:

$$p_{Cooler} = p_a + p_{BC}. \tag{1}$$

All data was time-averaged for each operating point, using an arithmetic average over all measured samples. With the described measurement data, the thermodynamic state of the cooling air within the model generator is well defined, and the cold air reference density was derived for each operating condition according to the ISO 5801 standard [1].

3.3 Volumetric flowrate

The flowrate of the cooling air was measured at the inlet of each of both fans. A Venturi type measurement principle was used, where the wall pressures for two positions with different cross-sectional areas at the fan inlet nozzle were measured and its pressure difference was correlated to the present flowrate.

The correlation of pressure difference and flowrate was done by an in situ calibration at the model generator. To do so, a calibrated measuring section was

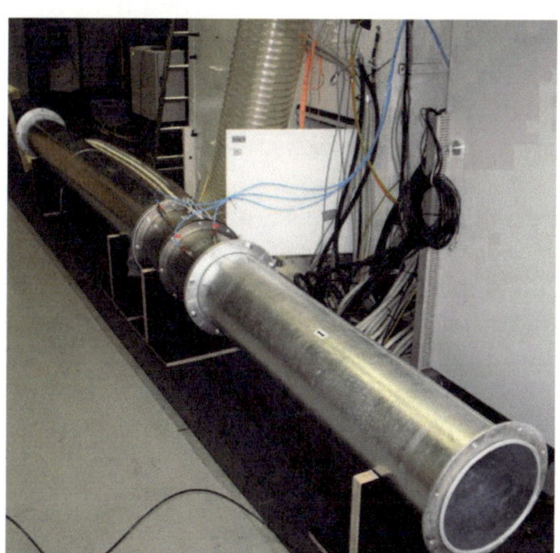

Figure 6.
Volumetric flowrate calibration measurement section.

installed on the pressure side of each fan. The measurement section can be seen in **Figure 6**; it is connected to the fan via an air hose on the far end side in the picture. With this setup, calibration data was obtained for the full operating range of the fans in their actual assembly situation.

From the obtained calibration data, a formula was derived to directly correlate the pressure difference measurement at the fan inlets and the volumetric flowrate in the measurement section. In accordance with the theory for an incompressible, Venturi type flow meter [2], the flowrate $\dot{V}_{Fan}$ through each fan is defined by

$$\dot{V}_{Fan} = a \cdot \left(\Delta p_{Fan,Venturi} \cdot \frac{\rho_{ref}}{\rho_{Cooler}} \right)^{b}. \tag{2}$$

In this equation $\Delta p_{Fan,Venturi}$ is the relative pressure difference at the fan inlet nozzle, ρ_{ref} is the air density during in situ calibration, and ρ_{Cooler} is the density obtained for each measured operating point behind the air-to-water cooler (Section 3.2). Factors a and b are the calibration coefficients derived from the in situ calibration measurement data.

4. Simulation setup

In the following sections, the simulation model is described in detail. An overview is given for the modelled geometry and the computational domain. The numerical mesh is presented. Also, the applied numerical models are described, and an overview is given about the different investigated numerical setups.

4.1 Model overview

The simulation model included all relevant generator parts modelled in high geometrical detail. Due to the single cooler on the top of the test rig, no circumferentially periodic model segment could be derived as recommended in [3]. The CFD model featured the complete geometry of the machine. Though this led to high numerical effort, a direct comparison with all available measurement locations was possible.

Instead of modelling the fans, the measured volume flowrate was specified for the respective operating point. The cooler was modelled as porous media, where the pressure loss was taken into account by an additional momentum sink term in the equations being solved during the CFD simulation. The underlying porosity model (momentum sink) applies parameters which were derived from the pressure measurements.

An overview of the model is given in **Figure 7**. **Figure 8** shows a cut through the model, which allows a view on the rotor parts. Furthermore, the geometrically modelled perforated metal plates can be seen, which produce a uniform inflow into the cold air compartment. The walls of the parts form the boundaries for the fluid domains which were used for the CFD calculation. All in all there were three domains: one domain for the rotating parts, one for the stationary parts, and one for the porous cooler. All domains were connected by interfaces in order to form a coherent CFD simulation domain.

4.2 Meshing

A computational mesh was generated for each domain. The unstructured mesh consisted of polyhedral elements and two wall prism layers. The interfaces between

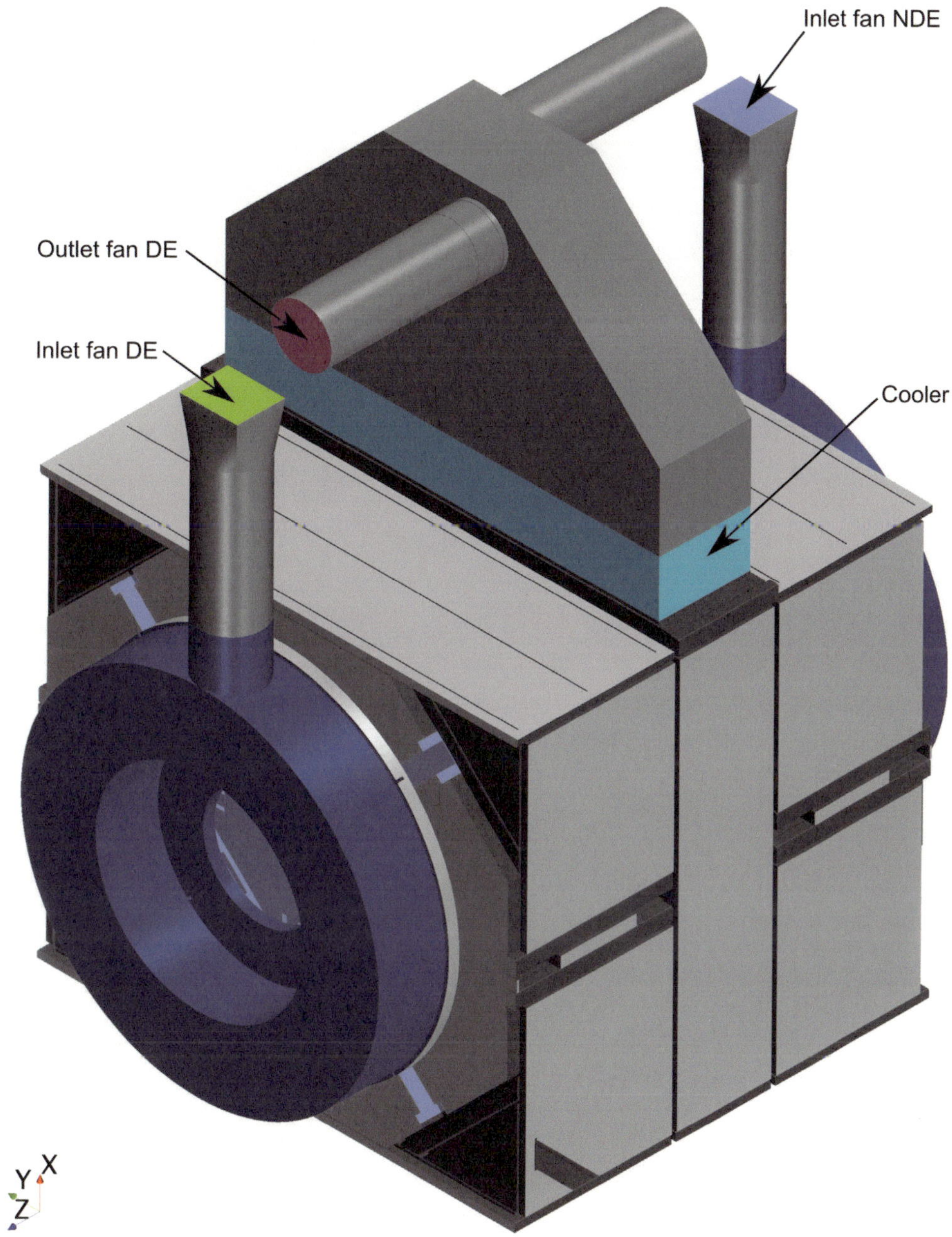

Figure 7.
Simulation model overview.

the domains were meshed node-conformal in order to improve the accuracy of the calculation. The total number of cells was 206.6 million.

In **Figure 9**, a section of the mesh within the pole gap can be seen. Inside the pole gap, the air cools the pole winding. To increase the heat transferring surface area, the pole winding is equipped with triangularly shaped cooling fins. To resolve this geometry feature, a fine mesh is mandatory.

Figure 10 shows the surface mesh of the rotor-stator interface, the cell size in the air gap between the interface and stator core, and one cooling channel. It can be seen that the mesh size is small in the area of the air gap and the intake of the cooling channels to resolve the high velocity gradients that are present in this area.

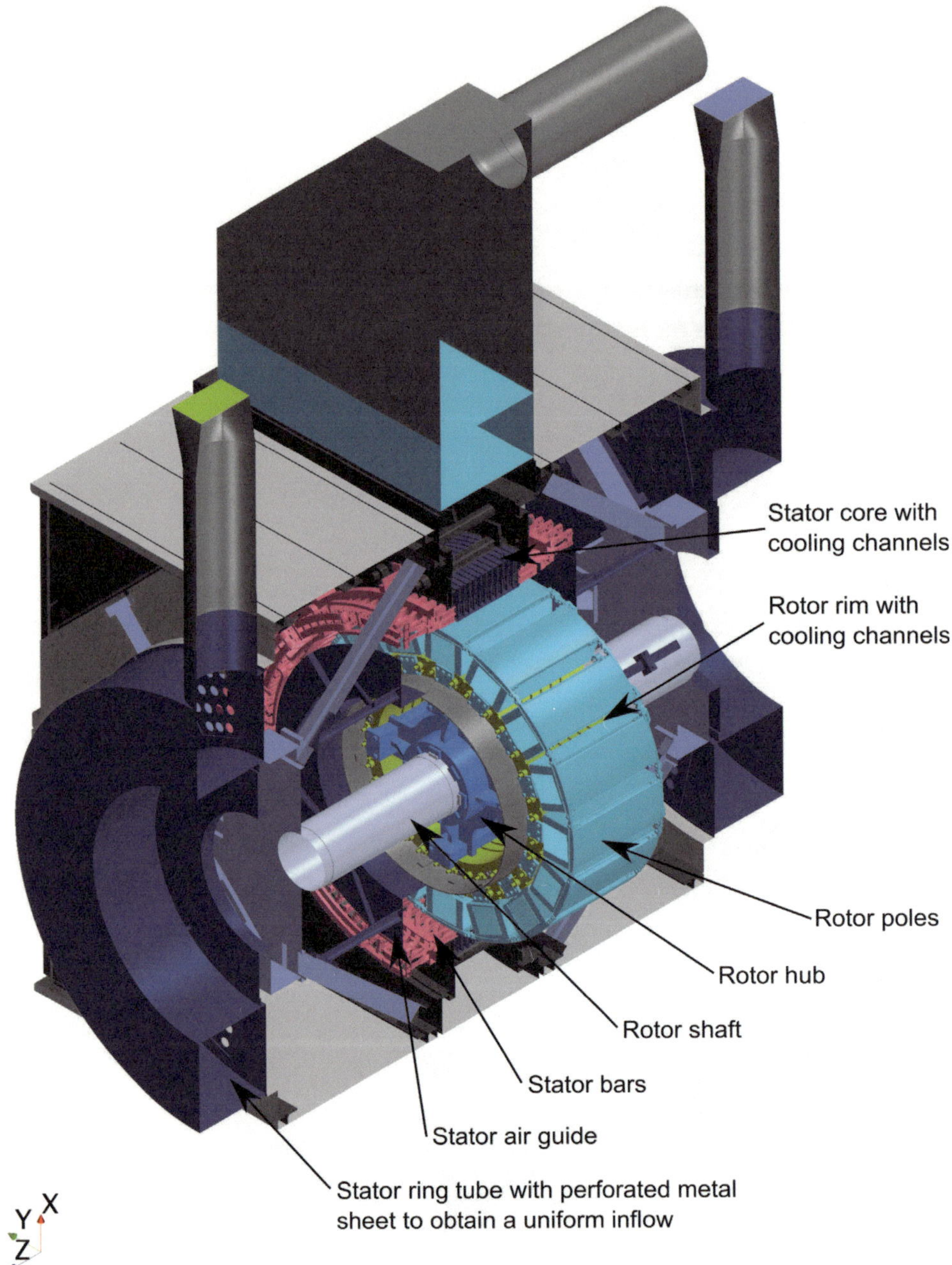

Figure 8.
Simulation model overview. Cut through stationary parts.

After entering the cooling channel, the airflow is guided in radial direction and is expected to develop a boundary layer similar to a turbulent channel flow with a low radial pressure gradient. This is why the cell size can be increased in radial direction in order to reduce the total number of cells.

4.3 Physical setup

An overview of the numerical setup is given in **Table 3**. All simulations were done using the segregated solver in the commercial software STAR-CCM+ v11.06. Steady-state RANS simulations [4] were conducted. The fluid was modelled with a

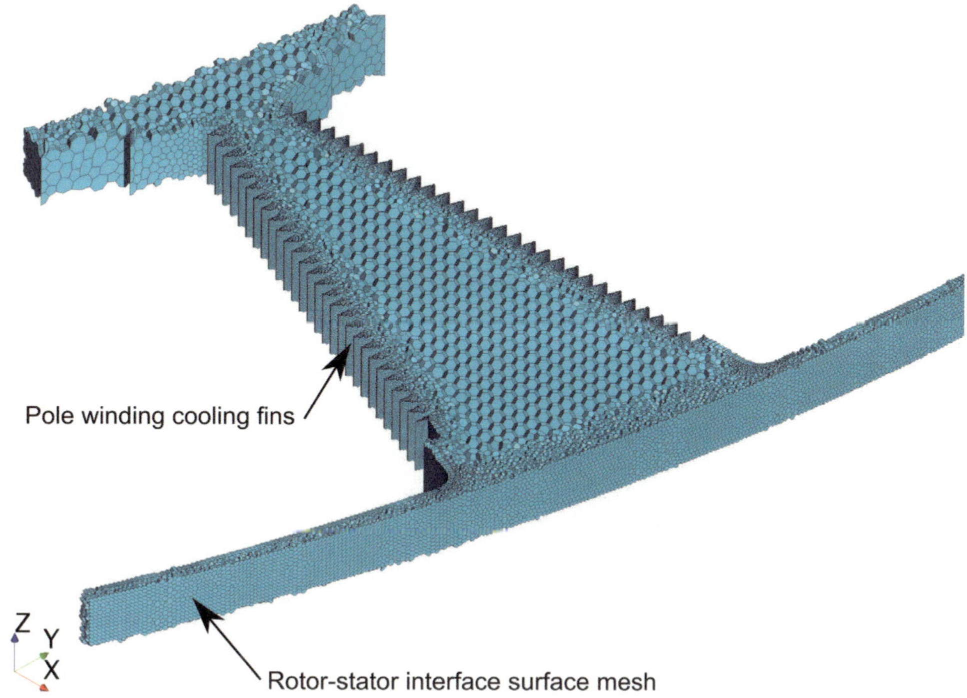

Figure 9.
Mesh overview. Pole gap.

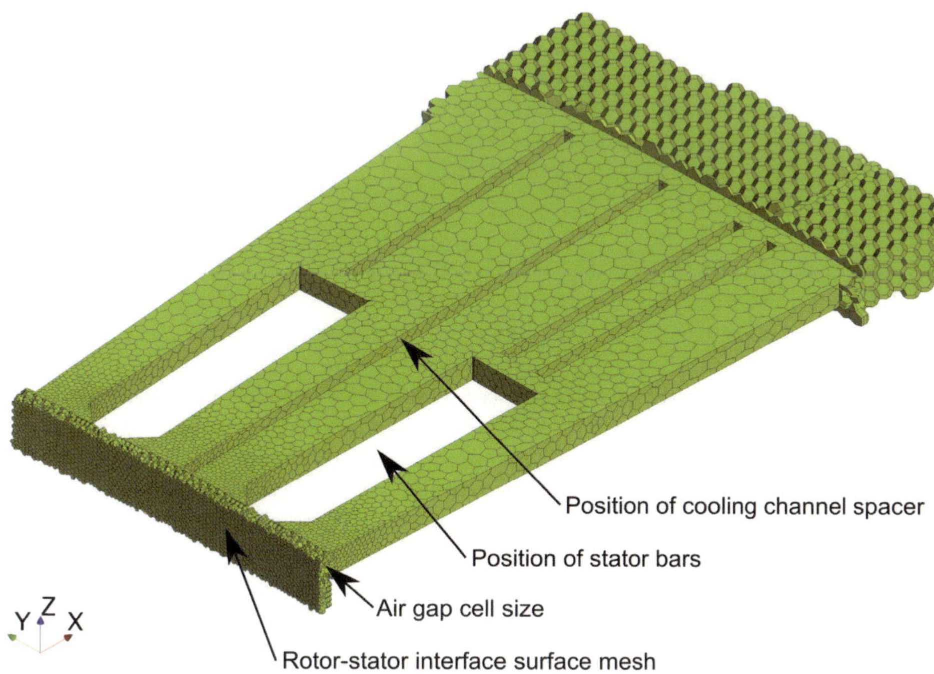

Figure 10.
Mesh overview. Stator cooling channels.

constant user-specified density and was considered isothermal. The density and dynamic viscosity were derived from the measured values (Section 3.2). The steady-state multiple reference frame approach was used for the simulation.

Solver	STAR-CCM+ v11.06
Analysis type	Steady-state
Rotor-stator interface	Frozen rotor/Mixing plane
Fluid model	Constant density
Heat transfer	Isothermal
Turbulence models	k-ε/k-ω SST

Table 3.
Physical setup overview.

Different modelling methodologies were investigated, including the influence of the frozen rotor and mixing plane rotor-stator interfaces and the effect of the k-ε [5] and k-ω SST [6] turbulence models.

Using the frozen rotor interface, the solution for one relative position of rotor and stator is calculated. The mixing plane interface leads to a circumferential averaging of the calculated flow quantities.

The k-ω SST turbulence model combines the benefits of the k-ε and k-ω turbulence model through blending. The k-ε turbulence model leads to better simulation results in free flows, whereas the k-ω turbulence model is expected to provide better turbulence modelling for near-wall flows [7].

5. Result evaluation

In the following sections, the CFD results are compared with the obtained measurement data. The pressure data for different positions in the machine is evaluated for a range of operating conditions. The focus of the evaluation is to identify coincidence as well as differences between the measurement data and the CFD simulation results with different numerical setups.

5.1 Flow path diagram

Figure 11 shows the averaged static pressure for each measurement position when following the flow path through the ventilation circuit. The values are normalized by the maximum fan outlet pressure that was measured for all operating conditions. The positions are named according to **Table 2**. For each flow path position, the value in the diagram is the average of all associated measurement positions that are available, e.g., CA 1–4 for the cold air compartment. The diagram shows the pressure plots for the four different fan speeds at a machine speed of 750 rpm. The other machine speeds have also been investigated but have shown qualitatively similar results, which is why only the 750 rpm variant is discussed here as representative example.

When the air passes through the fans, a static pressure rise can be observed for all operating points except for the point with fans turned off. In this case, consequently, a small pressure drop was measured. From fan outlet to cold air compartment, a pressure decrease can be observed, which is caused by a total pressure drop especially across the perforated plates mounted at the inlet of the cold air compartment.

In the cold air compartment, the static pressure of all operating points shown in the diagram approaches a value of zero. The reason for this is that in this area, there are several leaks to the outside of the machine, e.g., via the rotor shaft sealing. Therefore, the pressure equalizes with the ambient state in the laboratory.

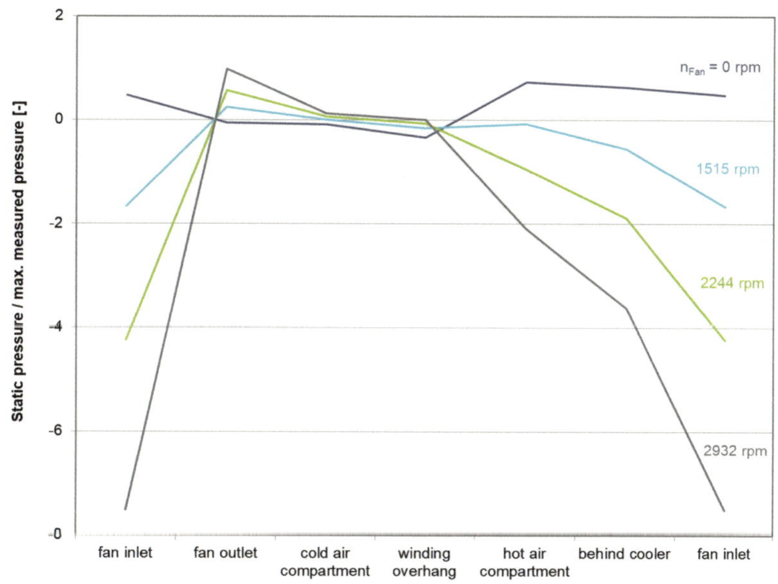

Figure 11.
Measured flow path diagram for $n_{Machine}$ = 750 rpm. Pressure values normalized by the maximum measured pressure.

Only a small pressure loss is observed from the cold air compartment to the winding overhang region for all operating points investigated. For the pressure drop from winding overhang to the hot air compartment, different effects can be seen, depending on the operating conditions.

With the fans turned off and at 1515 rpm fan speed, a pressure increase was measured. This is because the rotating components of the machine, comparable to a pump, transfer energy into the fluid which increases the total pressure of the system. In parallel, the flowrate provided by the external fans is rather low, leading to a low pressure loss within the ventilation circuit. Therefore, the pressure increase dominates over the pressure loss for the mentioned operating points. For the other operating conditions, the pressure loss in the machine is higher than its pressure buildup, as the external fans provide higher flowrates with increasing rotational fan speeds.

From the hot air compartment to the measurement position behind the cooler, a pressure drop can be observed for all operating conditions.

From the cooler outlet to the fan inlet, a comparably high pressure loss is present. This is due to the superposition of two effects: the total pressure loss along the flow path as well as a shift from static pressure to dynamic pressure as the cross-sectional area decreases when entering the duct toward the suction side of the fan.

The described pressure path is used to investigate the effect of different rotor-stator interface formulations and turbulence models. **Figure 12** shows all investigated combinations of rotor-stator interfaces (frozen rotor and mixing plane) and turbulence models (k-ε and k-ω SST) exemplary for the operating point with the highest machine and fan speed. The pressure level of the simulation data is adapted in order to obtain the same fan outlet pressure as in the measurements. The fan outlet is the inlet boundary of the CFD simulation model.

In **Figure 12** it can be seen that the calculated pressure drop between fan outlet and cold air compartment is approximately the same for all investigated combinations. Due to the fact that no rotor-stator interface is crossed, it can be stated that both turbulence models calculate the pressure drop for a combination of a duct flow

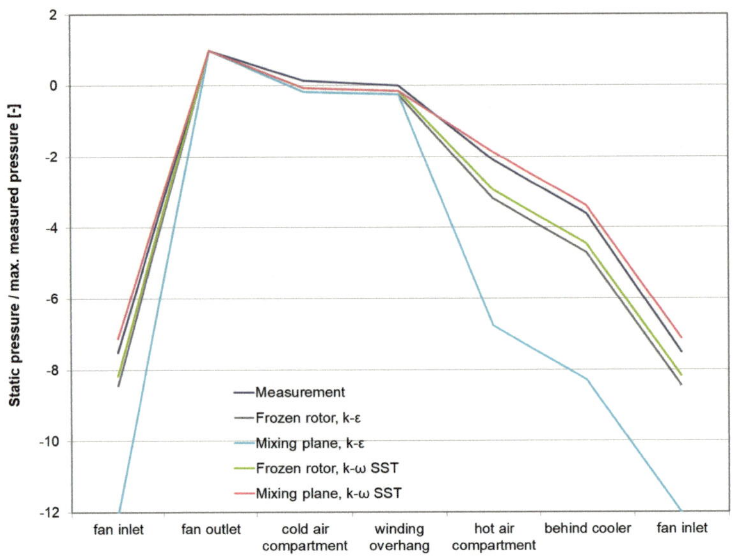

Figure 12.
Flow path diagram for $n_{Machine}$ = 750 rpm and n_{Fan} = 2932 rpm. Pressure values normalized by the maximum measured pressure.

and a geometrically modelled perforated plate in a stationary domain equally on the given mesh.

The major differences can be seen between winding overhang and hot air compartment. This is due to two effects. On the one hand, the influence of the rotor-stator interface can be observed, and on the other hand, the highest velocities and velocity gradients are present in this part of the machine. Therefore the influence of both, the rotor-stator interface and the turbulence model, can be seen.

For all simulations except for the combination of mixing plane interface and k-ε turbulence model, a comparable pressure drop is found. Both interfaces combined with the k-ε turbulence model calculate a rather high pressure drop. However, the combination of mixing plane and k-ε turbulence model overestimates the pressure drop significantly. Both interfaces in combination with the k-ω SST model lead to a good prediction of the pressure drop in this area, with the mixing plane model leading to the best agreement between measurement and CFD for the given operating point.

After the cooler, the pressure drop between cooler outlet and fan inlet is again comparable for all combinations.

Due to the high pressure drop for the combination of mixing plane interface and k-ε turbulence model between winding overhang and hot air compartment, the following evaluations were done for the k-ω SST model only.

5.2 Fan performance

The performance map of the ventilation circuit can be derived from the measurement data by plotting the pressure rise at the fans over the flowrate through the machine. As the static pressure before and after the fans was measured at positions with the same cross-sectional area and under the assumption that the velocity profiles in both positions are similar, it can be assumed that the static pressure rise is equal to the total pressure rise. **Figure 13** shows the performance map of the machine, with the static fan pressure rise Δp_{Fan} calculated according to **Table 2** as

$$\Delta p_{Fan} = \frac{1}{2}\left(\left(p_{FO\,1} - p_{FI\,1}\right) + \left(p_{FO\,2} - p_{FI\,2}\right)\right). \tag{3}$$

The differences in pressure rise and flowrates between both fans are small which justifies the averaging that is done in Eq. (3). The values in **Figure 13** are normalized by the maximum measured value for each fan pressure rise and volumetric flowrate.

The colored lines of constant machine rotational speed can be interpreted as consumer characteristic of the test rig, and the gray lines of constant fan rotational speed characterize the provider side of the setup.

With the fans turned off, a positive flowrate can be observed for machine speeds of 375 and 750 rpm. This is due to the rotating parts of the machine (pole, rim, hub) working on the fluid. The corresponding operating range is called self-ventilation.

When increasing the fan speed, the influence of self-ventilation is reduced, and the pressure rise due to the fan becomes the dominating flowrate source in the ventilation circuit. Therefore, the performance map shrinks to a rather narrow band, especially with comparably small differences between the consumer lines for 0 and 375 rpm. For the present machine setup, the maximum flowrate is reached at 750 rpm machine speed and 2932 rpm fan speed.

In **Figure 14** the performance map calculated from CFD with frozen rotor interface is shown. **Figure 15** shows the results for the mixing plane interface. Again, both diagrams are normalized by the maximum pressure and flowrate obtained from the measurements. To allow a direct comparison, the measured performance map is indicated by dashed gray lines.

In general it can be stated that the results for both interface models qualitatively agree well with the measurement data. All effects described above are represented in the simulation results, including the negative pressure rise in the area of low flowrates and the narrowing of the consumer lines at higher flows.

When concentrating on the machine speeds 0 and 375 rpm, also good quantitative agreement is reached between simulations and measurement for both interface formulations. The mixing plane results match the measured performance map well,

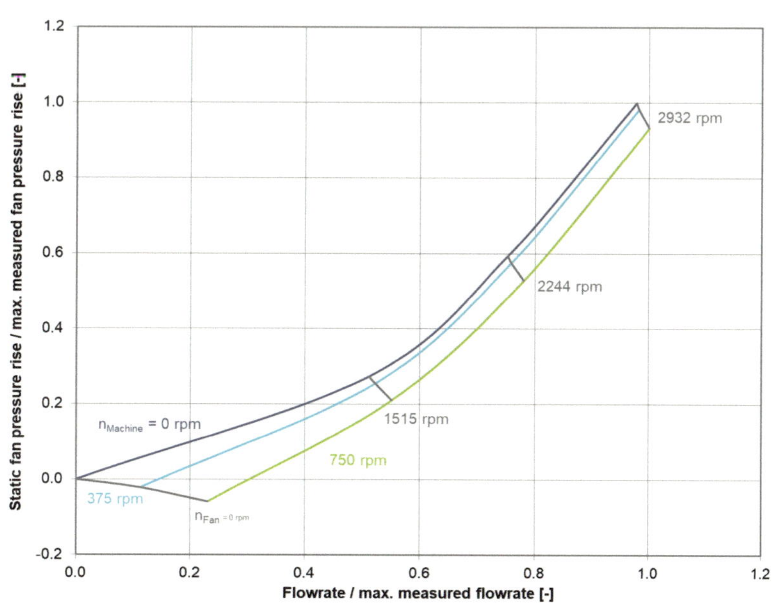

Figure 13.
Measured fan performance map. Values normalized by the maximum measured fan pressure rise and the maximum measured flowrate.

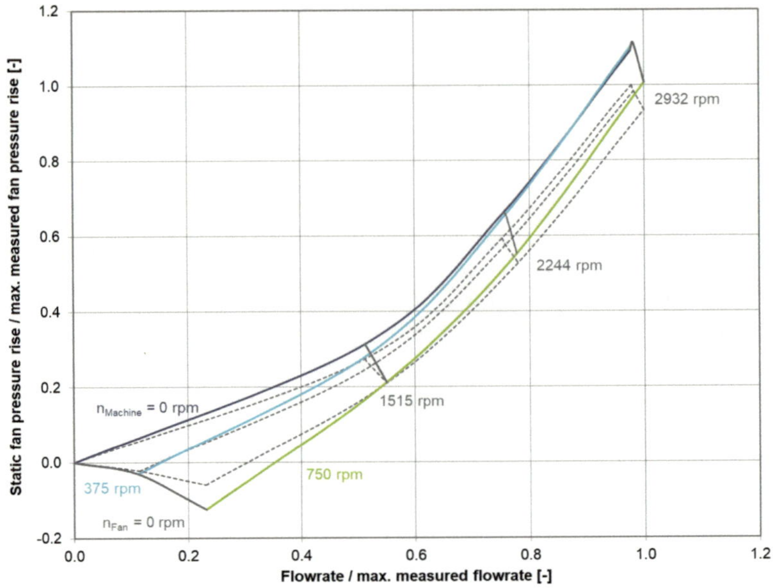

Figure 14.
Fan performance map calculated from CFD with frozen rotor interface. Thin dashed lines indicate the measured performance map for comparison. Values normalized by the maximum measured fan pressure rise and the maximum measured flowrate.

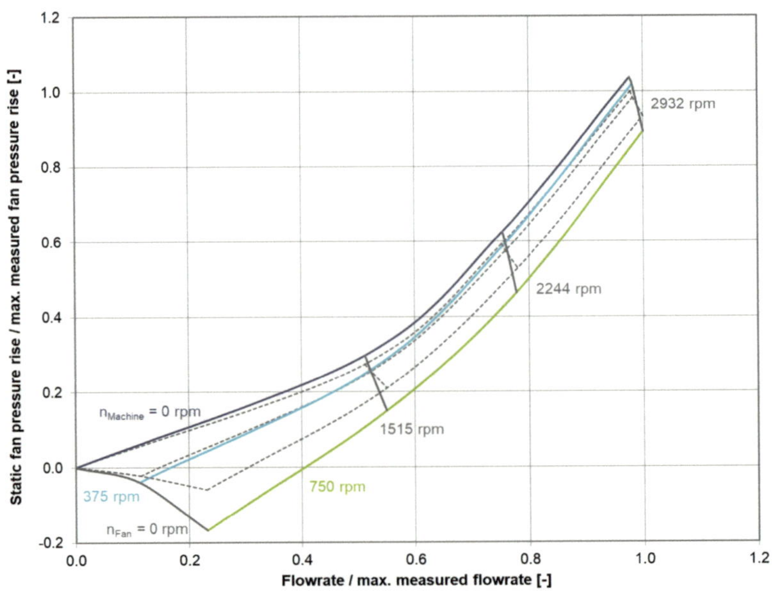

Figure 15.
Fan performance map calculated from CFD with mixing plane interface. Thin dashed lines indicate the measured performance map for comparison. Values normalized by the maximum measured fan pressure rise and the maximum measured flowrate.

especially in the area of higher flowrates. In this area, the frozen rotor interface tends to overestimate the machine's pressure loss.

Rather significant differences can be found at 750 rpm machine speed, particularly in the self-ventilation area with low flowrates. In this operating range, both interface models overestimate the negative pressure rise that is provided by the rotating electrical machine itself. This can be seen especially for the mixing plane interface where the negative pressure rise in the case of the fans turned off is predicted roughly two times higher than measured. This is an important finding as it shows that the simulation model reveals some sensitivity in predicting absolute pressure values when rotational effects dominate over the pressure loss of stationary parts in the system.

To evaluate the different predictions of such rotational effects, a closer look at the simulation results for 750 rpm machine speed and fans turned off shall be taken in the following. **Figure 16** shows static pressure plots for both interface definitions through the axial center of the machine. The pressure values are normalized by the magnitude of the measured fan pressure rise at the investigated operating point. The rotor pole gap, the air gap where the rotor-stator interface is positioned, and the stator core ventilation ducts can be seen in the picture.

First of all it can be noted that both contour plots appear very similar, with a qualitatively comparable pressure distribution. For both simulations, the absolute pressure difference depicted in the plot is higher than 14 times the magnitude of the measured fan pressure rise. Compared to this range, the differences in the predicted static fan pressure rise for frozen rotor (**Figure 14**) and mixing plane (**Figure 15**) which is similar to the absolute value of the measured fan pressure rise are rather small.

The most significant difference between both contour plots is the pressure level on the radial outside of the pole gap. Locally restrained to this circumferential position, the pressure is visibly higher for frozen rotor than for mixing plane.

The reason for this is that for the mixing plane interface, circumferential averaging of pressure and velocity leads to a well-distributed flow in the stator core ventilation ducts. Compared to this, for the frozen rotor interface, an increased amount of airflow is directly guided into the ventilation ducts at the same

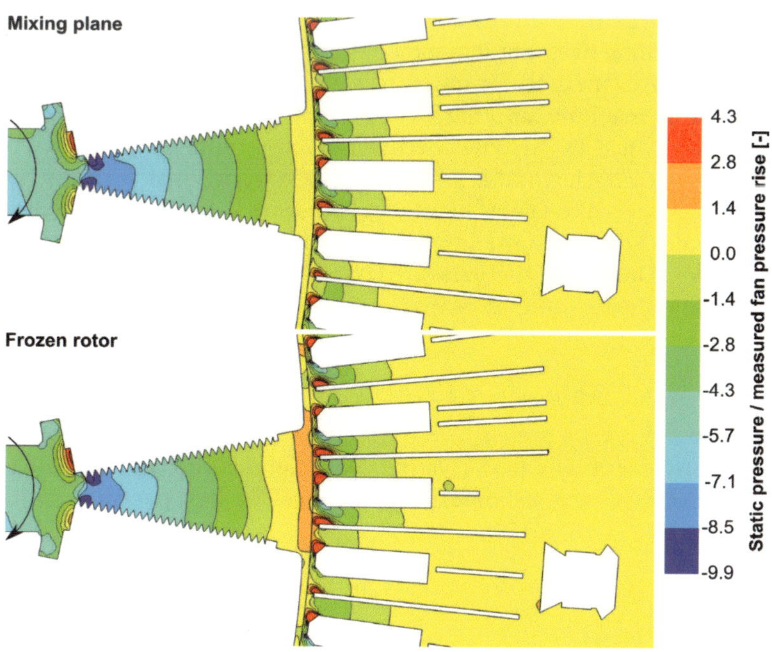

Figure 16.
Static pressure contour plot through the axial center of machine. Rotor pole gap, air gap, and stator core ventilation ducts are visible. Operating point is 750 rpm machine speed and fans turned off. Mixing plane (top) and frozen rotor (bottom) interface. Values normalized by the magnitude of the measured fan pressure rise at the investigated operating point.

circumferential position as the pole gap. Therefore, with the total mass flow being determined by the model boundary conditions, the pressure loss locally increases in the stator ducts in this position. In contrast to this effect, the circumferentially even flow distribution for the mixing plane interface leads to a comparably smaller overall pressure drop over the stator core ventilation ducts.

The described local increase of pressure radially outside of the pole gap for the frozen rotor interface propagates in upstream direction through the pole gap and eventually leads to the previously described differences in the performance maps of **Figures 14** and **15**.

5.3 Dimensionless machine parameters

It is a common practice to design hydro generator ventilation systems using analytical evaluations based on dimensionless parameters. Thus, in this section the capability to predict such dimensionless parameters shall be investigated for the CFD models. An approach to describe the pressure loss of a generator by three coefficients that can be assumed to be constant for geometrically similar machines is presented in [8]. In this method, a pressure loss coefficient ζ that is assumed to be proportional with the flowrate to the power of two is calculated for the machine in standstill. This yields the formula

$$\Delta p_{M,stat}(n,\dot{V}) = \Delta p_M(n=0,\dot{V}) = \rho\zeta\frac{\dot{V}^2}{{D_I}^4}, \tag{4}$$

where D_I is the inner diameter of the stator, n the rotational speed of the electric machine, and $\dot{V}$ the volumetric flowrate of the cooling circuit. The equation above shows a quadratic dependency with $\dot{V}$, whereas ζ is the only unknown. Therefore, the coefficient ζ can be derived from fitting this quadratic function to the measured and calculated machine pressure loss data in generator standstill. This was done using the least squares method for measurement and CFD.

For the rotating machine, another two coefficients that contribute to the pressure balance are defined in [8]. The coefficient ε represents an additional pressure loss due to machine rotation and is assumed to be proportional with the flowrate and the rotational speed; the coefficient ψ_0 represents the pressure buildup due to the rotating runner and is assumed to be proportional with the rotational speed to the power of two. Therefore, the pressure contribution by the rotating machine can be described as

$$\Delta p_{M,rot}(n,\dot{V}) = \rho\left[\varepsilon\frac{\dot{V}n}{D_I} + \psi_0 n^2 {D_I}^2\right]. \tag{5}$$

The most convenient way to derive the coefficients ε and ψ_0 from the measured data is to introduce two dimensionless parameters. The first parameter is the pressure number

$$\psi = \frac{\Delta p}{\rho n^2 {D_I}^2} \tag{6}$$

and the second parameter is the volume number

$$\varphi = \frac{\dot{V}}{n{D_I}^3}. \tag{7}$$

Inserting Eqs. (6) and (7) in (5) yields the dimensionless formula:

$$\psi_{M,rot}(\varphi) = \varepsilon\varphi + \psi_0. \tag{8}$$

Note that this equation is only valid for the rotating machine but becomes undefined for the generator in standstill. The advantage of the dimensionless form is that it only depends on one variable which is the volume number φ. By fitting the linear right hand side of Eq. (8) to the measured and calculated data using a least squares approach, the coefficients ε and ψ_0 can be derived.

The total machine pressure loss is the sum of the stationary and the rotating pressure loss. With Eqs. (4), (7), and (8), the machine pressure number can be described as

$$\psi_M(\varphi) = \psi_{M,stat}(\varphi) + \psi_{M,rot}(\varphi) = \zeta\varphi^2 + \varepsilon\varphi + \psi_0. \tag{9}$$

For the given machine, the influences of ζ and ψ_0 are dominating the pressure loss, whereas ε only has a small influence within the investigated operating range.

Figure 17 shows the pressure number ψ_M over the volume number φ for the measurement and the CFD simulations with frozen rotor and mixing plane interface. At lower volume numbers, both simulations underestimate the pressure number, whereas both simulations tend to overestimate the pressure number for higher volume numbers.

The mixing plane interface approximates the measurement results better for higher volume numbers and the frozen rotor interface for lower volume numbers. The evaluated operating points, excluding those without machine rotation, which are used to derive the necessary parameters described above, are also shown in **Figure 17** exemplarily for the simulation with mixing plane interface. These points help to understand the deviation between the two investigated interfaces. Most of the points for the highest machine rotation rate (blue crosses) correspond to lower volume numbers in comparison to the points for the mid rotation rate (gray crosses). Therefore, for this generator, the low-volume numbers are mainly

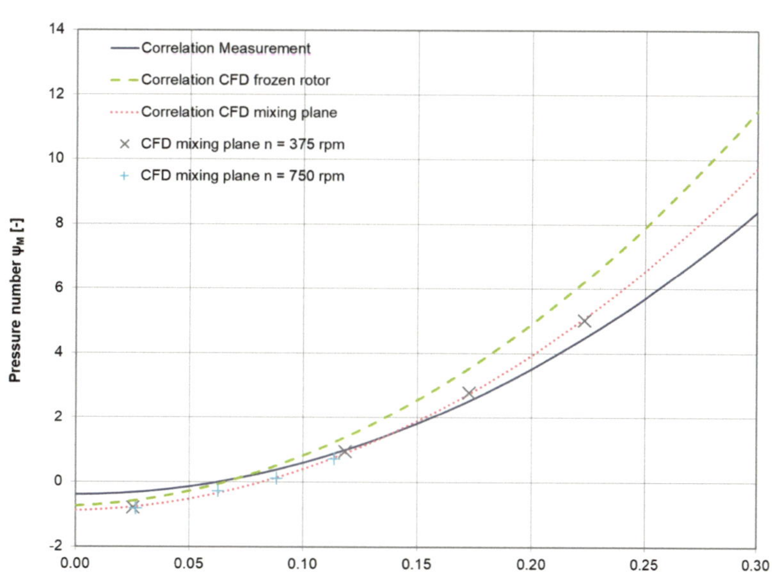

Figure 17.
Dimensionless machine pressure loss.

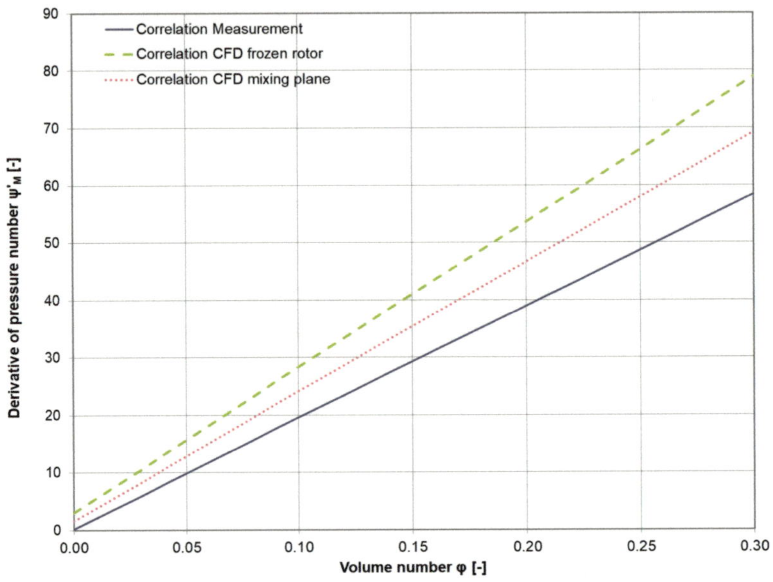

Figure 18.
Derivative of the dimensionless machine pressure loss.

influenced by the operating points with high rotational machine speeds. Comparable to Section 5.2, the frozen rotor interface leads to better results for the highest machine rotation speed, and therefore, the deviation between measurement and CFD simulation with frozen rotor interface is lower for low-volume numbers. For high-volume numbers, the influence of the medium machine speeds is higher, and therefore, the mixing plane interface results are closer to the measurement.

Figure 18 shows the derivative of the pressure number with respect to the volume number and therefore the gradients of the curves in **Figure 17**. The curves in **Figure 18** mainly show the influence of the pressure loss coefficient ζ (Eq. (9)). The lowest gradient can be observed for the measurements followed by the mixing plane interface and the frozen rotor interface. Therefore, the mixing plane interface approximates the gradient of the pressure number in dependency of the volume number to be more accurate in comparison to the frozen rotor interface, for the given machine configuration.

6. Summary and conclusions

Due to the need of detailed measurement data to validate simulation tools, Voith Hydro designed and built a model generator which allows the investigation of new generator design features or different configurations, e.g., different ventilation concepts. In this chapter, one specific ventilation concept was described.

Measurements and CFD simulations were conducted and compared to each other. The measurements showed the expected machine behavior and are therefore perfectly suitable to validate the presented simulation approaches. All in all, four simulation configurations were investigated which consist of two different rotor-stator interfaces and two turbulence models.

A comparison of the pressure paths showed that the combination of mixing plane interface and k-ε turbulence model overestimates the pressure drop between winding overhang and hot air compartment, whereas all other configurations showed a good prediction of the whole pressure path. The best approximation of the pressure path was reached with the mixing plane interface and the k-ω SST turbulence model, for the presented operating point.

For all other investigations, only the k-ω SST turbulence model was considered.

Looking at the fan performance map, good overall agreement between CFD and measurement was found for both investigated rotor-stator interfaces. However, there are several operating ranges where one of the interfaces is superior to the other one. A comparison of the dimensionless machine pressure loss showed that the curve gradients were resolved better with the mixing plane interface.

All in all it can be concluded that both combinations of rotor-stator interfaces and the k-ω SST turbulence model match the measurement data of the given ventilation concept well. This justifies the application of the presented simulation approach in the design of hydro power generator ventilation and cooling systems. Qualitatively, all measured flow effects were reproduced in the CFD simulations. When predicting absolute pressure levels, some deviations between measurement and CFD were present especially at lower flowrates, but in general good agreement was found over the full operating range.

As future work it is recommended to further investigate the effect of the chosen rotor-stator interface on the simulation accuracy for different operating conditions. In this context, transient simulations might be conducted for selected operating points. On the test rig, different ventilation concepts will be realized and investigated similarly to the measurement approach presented in this chapter. Furthermore, the model generator additionally is equipped with numerous temperature measurement sensors, providing validation data for even more detailed conjugate heat transfer simulations, which will also be investigated in future work.

Author details

Bastian Diebel, Axel Walter-Krause*, Roland Jester-Zuerker and Babette Schwarz
Voith Hydro Holding GmbH & Co. KG, Heidenheim, Germany

*Address all correspondence to: axel.walter-krause@voith.com

IntechOpen

© 2020 The Author(s). Licensee IntechOpen. Distributed under the terms of the Creative Commons Attribution - NonCommercial 4.0 License (https://creativecommons.org/licenses/by-nc/4.0/), which permits use, distribution and reproduction for non-commercial purposes, provided the original is properly cited. BY-NC

References

[1] ISO 5801:2017-09. Fans-Performance Testing Using Standardized Airways. Geneva, Switzerland: International Organization for Standardization; 2017

[2] Munson BR, Young DF, Okiishi TH. Fundamentals of Fluid Mechanics. J. Wiley & Sons: Hoboken, NJ; 2006

[3] Vogt B, Lahres S. Calculation of cooling and ventilation of large hydro generators with advanced 3D numerical simulations. In: Proceedings of the International HYDRO Conference; Austria. 2013

[4] Ferziger J, Peric M. Computational Methods for Fluid Dynamics. Berlin, Heidelberg: Springer; 2012

[5] Shih TH, Liou WW, Shabbir A, Yang Z, Zhu J. A new k-ε eddy viscosity model for high Reynolds number turbulent flows. Computers & Fluids. 1995;**24**(3):227-238

[6] Menter FR. Two-equation eddy-viscosity turbulence models for engineering applications. AIAA Journal. 1994;**32**(8):1598-1605

[7] Menter FR, Kuntz M, Langtry R. Ten years of industrial experience with the SST turbulence model. In: Hanjalić K, Nagano Y, Tummers MJ, editors. Turbulence, Heat and Mass Transfer 4. New York, NY: Begell House, Inc.; 2003. pp. 625-632

[8] Liebe W. Ueber das aerodynamische Verhalten elektrischer Maschinen. Siemens Zeitschrift. Berlin; 1957

Chapter 2

Hybrid Electro-Optic Capacitive Sensors for the Fault Diagnostic System of Hydrogenerator

Ievgen O. Zaitsev and Anatolii Levytskyi

Abstract

This chapter presents one of the ways of solving a problem ensuring reliability reducing instability and emergence of the power generators during its work. Its way use hybrid electro-optic sensors (HFOS) with capacitive mechanical sensors at their structure of the fault diagnosis system of hydro generators for measurement parameters of machine mechanical defects as quality parameters of air gap, shaft out, core compression ratio and other. So this chapter also contains: principle of measurement of the sensor, which based on measuring the mutual displacement of the flatness of the generator structural element relative to coplanar sensor electrodes surface; the design principle for developing HFOS with capacitive mechanical sensor which combines the benefits of microelectronic and fiber-optic technology; determination response characteristics of the sensitive sensor (capacitive sensors with coplanar electrodes) of HFOS for monitoring air gap defects and power accumulators of core clamping system and system of control; analytical calculations and experimental studies of air gap HFOS with the AD7746 at its operating excitation frequency using; optimum geometry calculation of air gap sensor electrodes for bulb hydro generators type SGK 538/160-70M; analyze application the air gap HFOS for the control system in the bulb hydro generators type SGK 538/160-70M.

Keywords: hydro generator, air gap, hybrid, measurement, capacitive sensor, displacement

1. Introduction

Hydroelectric power plants (HPP) are an important part of generating capacities that are capable not only for generating electricity but also due to the high mobility of hydraulic units and the pumped storage power plants use to provide a balanced load on the unified power grid during peak hours. Most of the hydroelectric power plants that are part Ukrhydroenergo PJSC have worked for a long time, more than 20 years. Only 45 hydraulic units from 103 were reconstructed in 2018, which made it possible to extend their lifespan by at least 40 years [1].

So, an important problem to be solved during the reconstruction of existing hydroelectric units as the design of new is to increase the reliability of operation and determine the actual technical condition of the hydraulic unit as a whole mechanical system. The solution to this problem requires a range of specialized measuring instruments for registration of diagnostic parameters, characterizing the flow of work processes in the hydraulic unit systems of their technical diagnostics.

Vibration diagnostics systems are widely used to obtain information on the state of the hydraulic unit. Sensors of this system mounted directly on the machine body for measured vibration parameters [2]. But, monitoring only the vibration parameters makes it possible to identify not all defects in various machine components. This is due with inaccuracy of diagnosis is often associated with measurement errors and incorrect interpretation of vibration measurement results, caused by insufficient temperature stability of vibration transducers, high noise levels, neglecting the features of machine vibration in the high-frequency and low-frequency areas, errors in the determination of discrete components of vibration during spectral analysis, and also due insufficient consideration in diagnostic systems specificity of stochastic fluctuations of the electromechanical equipment [3].

So, an important aspect to improve the performance of electric machines at the electricity generating station (thermal, atomic, hydro) it is the monitoring of their mechanical parameters. Change of mechanical parameters characterizes technical condition of the electric machine's equipment and exerts impact on the main energetic affectivities of working the generating station [4, 5].

Detection of defects which appear in the operating time of the machine at an initial stage of their emergence and timely acceptance the right decisions on their elimination before the emergence of an emergency situation provides a high level of readiness, reducing of downtime, lowering of costs of repairs.

Nowadays the usual ways to improve the performance of new and existing electrical equipment is design and realized new instruments for monitoring mechanical parameters of electrical equipment of the power generator [6–9].

2. Air gap as crucial parameters for characterizing technical condition of HPP

One of the major mechanical parameters that characterize the technical condition of electric machines equipment is its air gap. In hydro generators it is provided, the air gap is small in small as compared with the stator bore diameter. The air gap is a nominal or measured value between the hydrogenerator rotor and stator, or literally speaking it is the "heart" of a hydrogenerator, because in the air gap the mechanical energy is transformed into the electric energy [10]. While machine work it is difficult to obtain stable size and uniformity of the air gap. The air gap deviations from nominal value are resulting of distortion of the air gap between the stator core and rotor. The air gap deviations from nominal value are resulting of distortion of the air gap between the stator core and rotor. The air gap deviations may be caused by deviations of machine construction, static or dynamic eccentricities, cone rotor, ellipse-shaped surfaces of the stator and rotor and other factors [11]. Technological inaccuracies resulting in an eccentricity are practically impossible to avoid in a production process of the power generators for HPP [12, 13].

Therefore control and monitoring of the air gap that is one of the most important quality parameters of HPP is very important. For doing this usually way it is the air gap sensors use [6–9]. Information's from the sensors in automated measuring systems used to simplify testing and minimize repair time, to avoid accidents during machine operation. Capacitive sensors for measuring air gap in hydrogenerators are represented by the following companies:

- Vibro-Meter LS120 for range 2–33 mm [14]

- Iris Power for different ranges from 5 to 47 mm [15];

- Bently Nevada offers a 4000 Series Air Gap Sensor System with an air gap sensor up to 20 mm [16];
- HUBER+SUHNER has developed the air gap sensor consisting of a fiber-optic loop [17].

All existing sensors are designed to be used only on definite types of generators. Sensors adaptation underuse into a definite type generator requires considerable financial costs [16, 18]. So promising for the construction of technical diagnostics of power equipment is used HFOS [4, 16]. HFOS combines the benefits of fiber optic and microelectronic technologies [8, 9, 19].

3. Air gap parameters and their influence on key characteristics of hydrogenerators

The air gap between the rotor and the stator is one of the most important parameters of a powerful generator. The size of the air gap largely determines the characteristics of the machine and its behavior during operation.

Due to the irregularity of the air gap, an asymmetry of electromagnetic forces occurs in the GG, which in turn leads to the appearance of vibration and an increase in the surface temperature of the rotor. One of the signs that the vibration has arisen due to the uneven gap is its dependence on the excitation current. When the excitation is removed, the vibration disappears completely.

The small size of the pole distributions of the hydrogenerators causes large scattering of their pole system, high inductions in the cores of the poles and causes difficulty in placing the required volume of the excitation copper winding at the poles. This leads to hydrogenerators resort to reduce air gaps. In practice, given the above factors, the values of air gaps are taken in one thousandth of the rotor diameter.

The value of the air gap also significantly influences a number of other characteristics of hydrogenerators, namely: the end magnetic fluxes of losses in the extreme packets of the core and the stator pressure plates, the vibrational state of the machine during operation. Thus, both the thermal state of the hydrogenerators as a whole and the level of local heating are largely determined by the value of the air gap. In addition, the areas of acceptable modes of operation of hydrogenerators (with the condition of non-excitation) are also determined by the value of the air gap [20–30].

Distortion of the rotor and stator shape of the hydrogenerators, which leads to an uneven air gap, can cause accidents (due to rotor engagement on the stator), failure of the windings, steel cores, and poles. It is especially difficult to achieve a stable operating gap even if the air gap is small (compared to the stator bore diameter). Creating a design that resists the magnetic pull in such a machine configuration is one of the most difficult tasks in the development of hydrogenerators.

Often, damage caused by a change in shape or a violation of the alignment of the stator and rotor cores has been observed. Thus, the largest at one time HG HPP Grand Coulee Dam (USA), having an air gap of 25 mm with a rotor diameter of 19 m, repeatedly failed due to the displacement of the active parts of the machine. At the same time, there was even engagement of the rotor for the stator, which led to the need for their reconstruction with increasing the air gap and increasing the strength of the stator core [20]. Serious problems with deformation of the rim of the rotor were on the hydropower generators Mica Creek (Canada). These machines, with a diameter of 14 m, have a nominal air gap value of 20.6 mm [20].

The difficulties with maintaining the air gap within the acceptable limits for the hydrogenerator is illustrated by the example given by a representative of Alstom Jeumont [20]. When the stator core is heated, the diameter of 17 m will expand by 5 mm and the rotor by 5.3 mm (the estimated air gap for this machine is 24 mm). Considering also the magnetic pull forces between the stator and the rotor, which are aimed at reducing the air gap, the designer has a difficult task to maintain the cylindricality of the rotor rim, the uniform expansion of the stator and the spokes of the crosspiece.

During operation of capsule SGs type SGK 538/160-70 of Kyiv and Kaniv hydroelectric power stations, due to the distortion of the stator shape and imperfect rotor shape, the air gap between the rotor and stator varied in both radial and axial direction: at a nominal value of 5 mm the gap did not exceed 3.5 mm [24, 29]. Due to the significant reduction of the air gap, the load on the damper winding, which it was not designed for, increased sharply. In the end, this led to the destruction of the damper system: there was a rupture of the cores, the burning of core steel and the fall of the damper rods from the core into the air gap, as well as significant damage to the core and stator winding. During the reconstruction of these hydropower generators, the following was done to prevent such accidents: the nominal air gap between the rotor and the stator was increased to 6 mm, the adjusting gaskets for the rotor poles were introduced, and the requirements for the correct form of the stator bore were increased.

The irregularity of the air gap, accompanied by the deviation of the bore of the stator core and the bypass pole of the rotor from the cylindrical shape, can occur during the installation, after it, as well as under operating conditions of the generator. Electromagnetic forces of the mutual attraction of the stator and rotor, as well as redistribution of internal stresses, can lead to deformation of the stator and rotor core. Uneven heating of the stator packs and rotor poles can also contribute to the deformation of the generator units.

It is also known that the irregularity of the air gap in height and radius of the magnetic system of the hydrogenerator is the cause of the vibration of the stator and rotor and the cause of additional losses on the surface of the rotor poles [30].

The deviation of the air gap from the nominal value can also be caused by defects in the design of fastening elements, violations of the technology of assembly of the hydrogenerator at the station, degradation processes during operation of the unit under the action of electromagnetic and thermomechanical loads [20].

Analyzing the above, it can be argued that to control the air gap in powerful GH requires reliable high-performance devices, designed with the design features of the machine.

4. The architecture of the control system with HFOS for electric machine faults monitoring system

The architecture of electric machines faults diagnostic system incorporates a wide range of mechanical sensors to control many technological parameters. However, in practice, during operation of the machine, external influences (electromagnetic fields, temperature, etc.) act on the measuring equipment. For avoiding external influence on the sensor and control systems operation unit is used fiber optic and other electro-optical components. The benefits of the optics components make possible to use HFOS in applications for measurement parameters of mechanical faults in working environment of large generator with high-EMI field, temperature, pressure, explosive, noisy, etc. [9].

The HFOS structure for large generator faults monitoring system shows on **Figure 1**, has following notation: PSCS – primary special capacitive sensor of the mechanical parameters with capacitance-to-digital converter on board; SigPr – microcontroller unit or signal processing unit; SW – switch; OPS – converter optical energy to power supply of HFOS unit; OF – optical fiber; OCS CMP – control systems operation unit for large generator fault diagnostic system; Optical Tx/Rx – indirect way its converter optical digital code to electrical digital code with digital controller and reversed way its electrical to optical digital code.

Figure 1 shows a simplified structure of control system processing tools with a complete mapping of the structure of the HFOS microelectronic and optoelectronic components in functional groups. The system consists of two parts: HFOS in isolated area and processing tools of the monitoring system. The monitoring system processing tools located in isolated areas on safe distance which provides a low level of external influences (electromagnetic fields, temperature, etc.) [8, 9].

Group of unit of HFOS in isolated area integrates a capacitance-to-digital converter (CDC) AD7745/AD7746 from Analog Devices, Inc. (Norwood, MA, USA) [31] on primary capacitive sensor circuit board material made from FR4 copper clad laminate printed circuit board (PCB) [32], optical Tx/Rx implementations by analogy described in [33] with low power microcontroller and vertical-cavity surface-emitting semiconductor laser with very low threshold current, semiconductor photovoltaic energy converter (labeled as the “OPS”).

In the system described in the work [9], a pair of optical fibers connects the OPS and vertical cavity surface emitting semiconductor laser to the external system to receive power energy by optical cable and measurement data by fiber optic. Moreover, one optical fiber to receive control data to microcontroller from main system used.

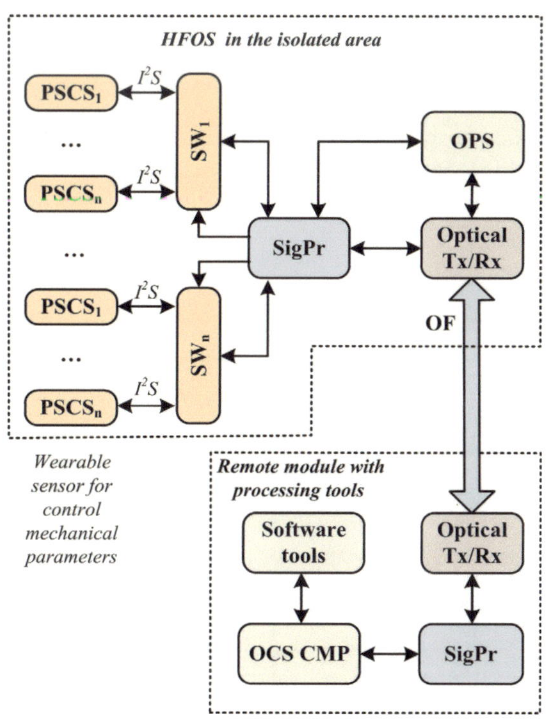

Figure 1.
Control system with HFOS.

The group of blocks in a safe area integrates "optical Tx/Rx," SigPr system and digital part with a special software solution for detection and monitoring mechanical faults of the electric machines. In the remote zone there is a user interface through which the operator controls the operation of the system and receives information about the state of the object.

5. Work principle by HFOS for large generator faults monitoring system

5.1 Principle of work primary capacitance sensor for electric machines mechanical faults monitoring system

Capacitive sensors can be applied for measuring a different kind of non-electrical quantities [34, 35], such as mechanical parameters of electric machines equipment: geometrical dimensions change, displacement and vibration of grounded surfaces, air gap, and position of the object, core compression ratio and others. Capacitive mechanical sensors are the most widely used non-optical sensors in short-range positioning applications owing to their excellent resolution [18].

The principle of measurement of a capacitive sensor is based on the change in the capacitance with the distance of the capacitor with parallel electrodes coplanar witch are located in one place. **Figure 2** has shown the typical design of coplanar sensor. The coplanar sensor consists of the following parts: hi-potential electrode 1; low-potential electrode 2; low-potential guarding electrode 3; dielectric substrate 4; metal substrate 5, which is installed on a grounded surface special elastic element which is located on the stator between stator bore and rotor polys [36].

Electrical capacity C of the coplanar sensor between electrodes 1 and 2 will change when change air gap between rotor and stator bore of the hydrogenerator [8] and can be calculated as [32, 36]:

$$C = f(d) \tag{1}$$

where d is a distance between the electrodes of the capacitance sensor.

Based on this principle was implemented mechanical sensor for measuring the value of air gap and other in the hydrogenerator.

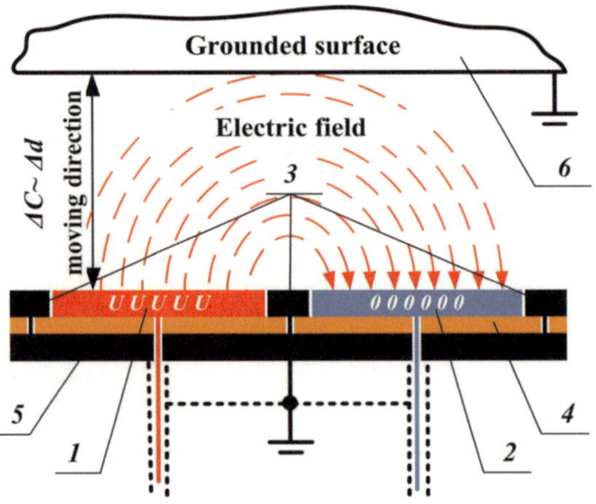

Figure 2.
Capacity sensor with coplanar electrodes.

5.2 Principle of work microelectronic components

The base of the principle of operation the system shown in **Figure 1** is described in [9] and is as follows: the primary measuring transducer (sensing element PSCS) converts the value of a controlled parameter of a mechanical defect into a digital code like NRZ. Then the SigPr communication subsystem collects measurement information from the PSCS and converts the information data into a modulated optical signal [9]. The signal is transmitted by optic fiber-cable (OF) to optical Tx/Rx module where optical signal converted into an electrical signal. The electrical signal as digital code like NRZ send to processing tools as data information processing system OCS CMP and use for analysis with special software tools. In the case of the transmission of digital control data signals for PSCS transducers, the OCS CMP system works in the similarly [9]. The architecture of HFOS was detail analysis in work [8, 9].

For converter capacitance value of the primary sensor to digital code chosen capacitance-to-digital converter (CDC) AD7745/AD7746 (Analog Devices, Inc., Norwood, MA, USA) with temperature sensor. The use of a 24-bit sigma-delta converter allows the resolution of the measurement range of an informative capacity of 4fF [31].

5.3 Principle of work power optoelectronics components

The benefits of fiber optics make it possible to use the capacitive sensor in applications for measurement mechanical parameters of power generator in their working environment with high-EMI and magnetic field, temperature, pressure, hazardous, explosive, noisy and etc. Data sent by data optic fiber. Power supply for primary measuring converters HFOS in an isolated area can transmit by the following ways [19, 33, 34, 37–45]:

a. with the autonomous power supply:

- using battery power directly adjacent to the meters;
- with the help of energy from power sources realized by technology "Energy Harvesting" and located directly adjacent to and / or component of the meters;

b. powered by a fiber optic line:

- energy transmitted through the power optical line, which can be used:
    - i. single-mode fiber;
    - ii. multimode fiber;
    - iii. optical harness;
- using the energy transmitted through the information-power optical line, implemented on the basis of "Wavelength-Division Multiplexing" technologies (multimode fiber).

In the system that we are reporting, the second option was used when light connected to one optical fiber with a unique photovoltaic converter on the one hand, and a powerful laser that was located in SigPr on the other was used to power it. For photovoltaic energy converter it proposed to use semiconductor photovoltaic cells based on crystalline GaAs.

5.4 Principle of work software solution

Remove module of the monitoring system consist of optical converters with microcontroller for communication and a personal computer, which functions under the control of specialized software. Its functions designed for [5]:

- the definition of the class of possible defects (the most important or most frequently encountered) that must be identified;
- selection of diagnostic signals available for measurement, and control points on the object under study;
- development of a mathematical model of the diagnostic object, the analysis of which allows substantiating possible diagnostic parameters;
- development of algorithms for obtaining numerical values of selected diagnostic parameters;
- construction of decisive rules for identifying and classifying defects; creation of means implementing certain steps of the diagnostic process from the selected measurement and diagnostic signals before making diagnostic solutions.

The system software (primary data acquisition and processing module) consists of the following parts: a data processing software module from CDC and CDC control, a microcontroller configuration module, a primary data processing module [46].

The CDC control module algorithm for converter capacitive value to code and receive the digital data to the monitoring system use algorithm from [31] for "AD7746," the data exchange for designed to provide the organization of data exchange between hardware and software monitoring system and CDC.

Operations of calculating the value of the mechanical parameters special software solution is used. Processing of the received data in the control module of mechanical parameters state of the control is used for the value of the mechanical parameter and analyzed its status and changes. In turn, obtained at the work of the module of mathematical processing and module of automatic control of the state of the electrical equipment node is transferred to the data storage organization module for database management based on the history of measurements. In this case, it is possible to create knowledge bases with diagnostic features, which depend on the value of the physical parameter that is control of the certain state of the power equipment.

The connection of the monitoring system and HFOS module in the isolated area is transfer with a developed communication protocol for the fiber-optic line communication. In addition, the module for input and output information SigPr is designed for organizing the exchange data between the modules of the system.

6. Sensitive elements of air gap HFOS for the hydrogenerator faults monitoring system

6.1 Basic principle of the air gap control in the large hydrogenerator

Nowadays the usual method for the designed and realized sensor for the air gap measurement in the powerful hydrogenerator is a capacitive method [4, 10, 47–49], now. When you use the method of capacitive sensors mounted on the bore of the stator core.

In this method, the electrical C capacitance sensor depends on the size of the air gap d and can be calculated by Eq. (1).

In [50] present variant when the two capacitive sensors are mounted on the same plane with the angle between them is 90 degrees. **Figure 3** shows example of the installation air gap sensor with the control system on the stator core of the bulb hydro generators type SGK 538/160-70M. **Figure 3** has the following notation: 1 – stator core; 2 – air gap capacitive sensor; 3 – connecting cable between the sensor and the converter; 4 – secondary transmitter; 5 – connecting cable between the converter and the computer; 6 – computer with processing tools.

6.2 Influence on the measurement accuracy of the air gap of the skew of the sensor electrode relative to the surface of the stator core

Influence on the measurement accuracy can be calculated by use obtaining results in the paper [49–51] by the equations

$$\Delta C_{12i} = \frac{\varepsilon_0 \varepsilon}{\pi} \Delta y_i \ln \frac{\left(\mathrm{th}\frac{\pi s}{4z_i} + \mathrm{th}\frac{\pi(s+2b)}{4z_i}\right)^2}{4\mathrm{th}\frac{\pi(s+2b)}{4z_i}\,\mathrm{th}\frac{\pi s}{4z_i}}, \tag{2}$$

where b – width of the electrodes 1 and 2; s – the distance between the electrodes 1 and 2; z_i – the distance between the plane and the plane of the strips **5** and **4** after the onset of warp (**Figure 4**).

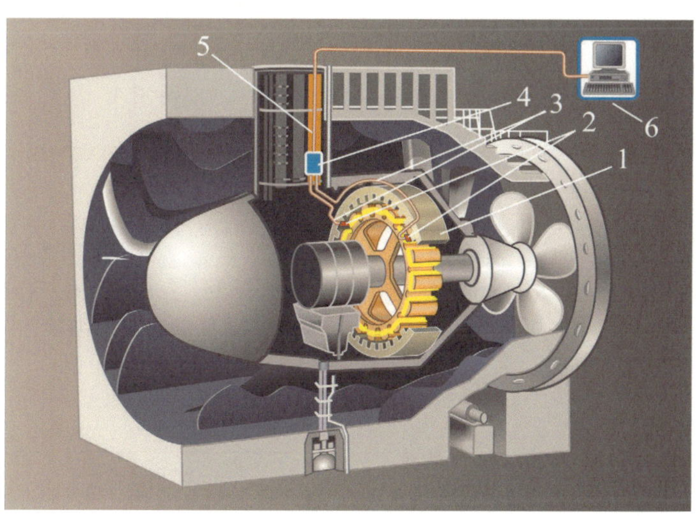

Figure 3.
Example of an installation control system with two HFOS.

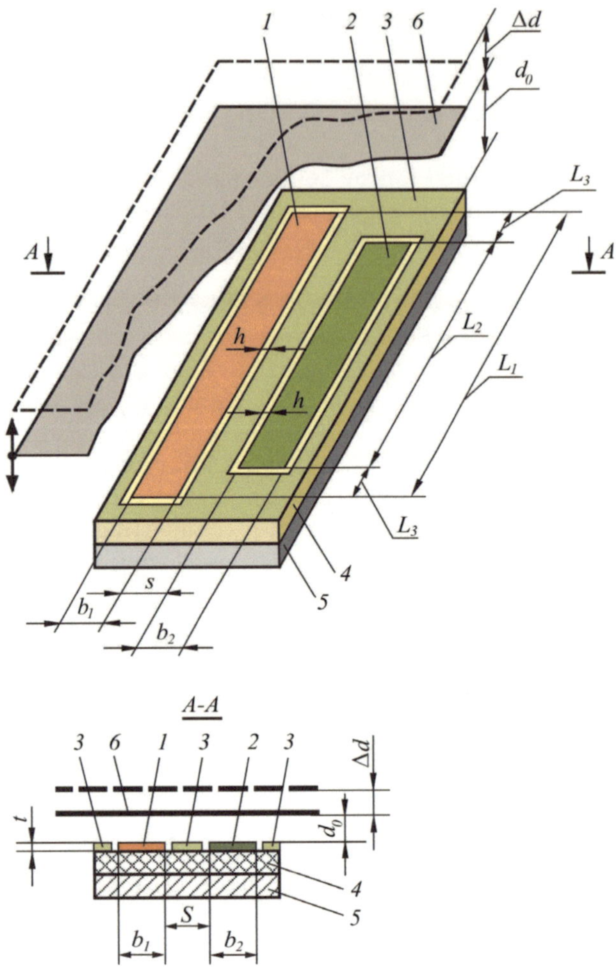

Figure 4.
Scheme for calculating the influence of plane electrode sensor.

Considering $A_0A_1 = d0$, $A_1A_2 = \Delta z_i$, $\Delta z_i/y_i = tg\alpha$, z_i defined as:

$$z_i = d_0 + \Delta z_i = d_0 + y_i\text{tg}\alpha, \tag{3}$$

where y_i – the distance between 5 and point of start coordinates O; d_0 – the distance between the plane and the plane of the electrodes 4 before warp; α – the angle between the plane and the plane of the electrodes 4, resulting warp.

The capacity $C_{12\Sigma}$ of the sensor use Eqs. (2) and (3) can calculate as:

$$C_{12\Sigma} = \sum_{i=1}^{i=\infty} C_{12i} = \sum_{i=1}^{i=\infty} \frac{\varepsilon_0\varepsilon}{\pi} \Delta y_i \ln \frac{\left(\text{th}\frac{\pi s}{4(d_0+y_i\text{tg}\alpha)} + \text{th}\frac{\pi(s+2b)}{4(d_0+y_i\text{tg}\alpha)}\right)^2}{4\text{th}\frac{\pi(s+2b)}{4(d_0+y_i\text{tg}\alpha)}\text{th}\frac{\pi s}{4(d_0+y_i\text{tg}\alpha)}}. \tag{4}$$

Without warp capacitance C_{12} between electrodes **1** and **2** is defined by Eq. (4) [50].

The error of warp δ_{Π} defined by using Eqs. (2) and (3) as:

$$\delta_{\Pi} = \frac{C_{12} - C_{12\Pi}}{C_{12}} \cdot 100\% = \left(1 - \frac{\frac{\varepsilon_0 \varepsilon}{\pi} \int_0^L \ln \frac{\left(\text{th} \frac{\pi s}{4(d_0 + y\text{tg}\alpha)} + \text{th} \frac{\pi(s+2b)}{4(d_0 + y\text{tg}\alpha)} \right)^2}{4\text{th} \frac{\pi(s+2b)}{4(d_0 + y\text{tg}\alpha)} \text{th} \frac{\pi s}{4(d_0 + y\text{tg}\alpha)}} dy}{L \frac{\varepsilon_0 \varepsilon_r}{\pi} \ln \frac{\left(\text{th} \frac{\pi s}{4d_0} + \text{th} \frac{\pi(s+2b)}{4d_0} \right)^2}{4\text{th} \frac{\pi(s \mid 2b)}{4d_0} \text{th} \frac{\pi s}{4d_0}}} \right) \cdot 100\% \tag{5}$$

Define capacity $C_{12\Sigma}$ and C_{12} sensor gap [50], for use in capsule hydrogenerators SGK 538/160-70M with next parameters b = 12 mm, s = 2 mm, L = 180 mm, using Eq. (4) and [32]. Result defines capacity $C_{12\Sigma}$ and C_{12} shown in **Figure 5**.

Using Eq. (7), determine the numerical value of warp error for the hydrogenerators SGK 538/160-70M and plot them in **Figures 6** and 7.

6.3 Experimental studies

6.3.1 Design of the sensitive elements of air gap HFOS

The air gap sensor for the control system in the hydrogenerator is designed and realized in the Department of electric and magnetic measurements of the Institute of Electrodynamics of the NAS of Ukraine. The development sensor with processing unit was shown in **Figure 8**.

6.3.2 Temperature stability

Generally, the coplanar air-gap sensor is developed and designed for work in temperature range from −30 to +80°C. The low-limit of temperature can be during

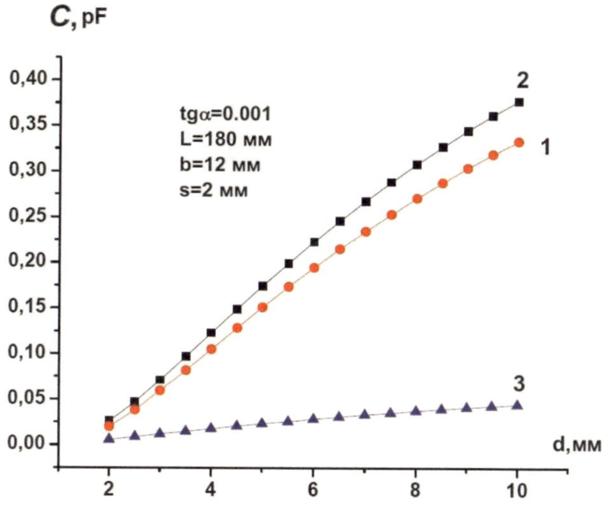

Figure 5.
Capacity $C_{12\Sigma}$ and C_{12} as measurement range: 1 – Change of capacitance $C_{12\Sigma}$ in a nominal measuring range from 2 to 10 mm with warp; 2 – change of capacity C12 without warp; 3 – the capacities difference $\Delta C = C_{12} - C_{12\Sigma}$.

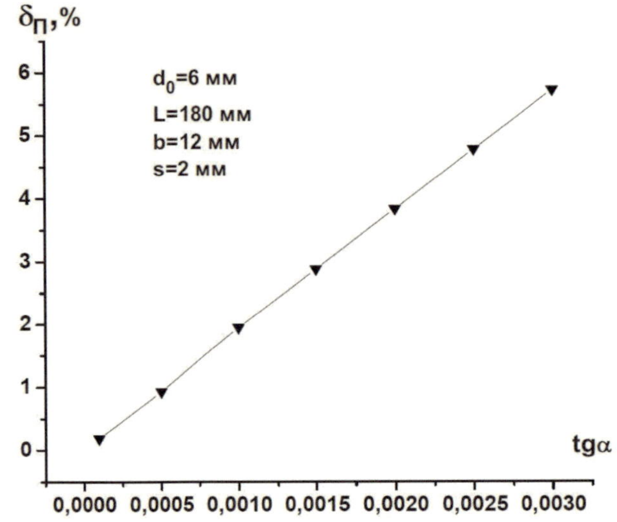

Figure 6.
Angle warp as error for d_o = 6 mm.

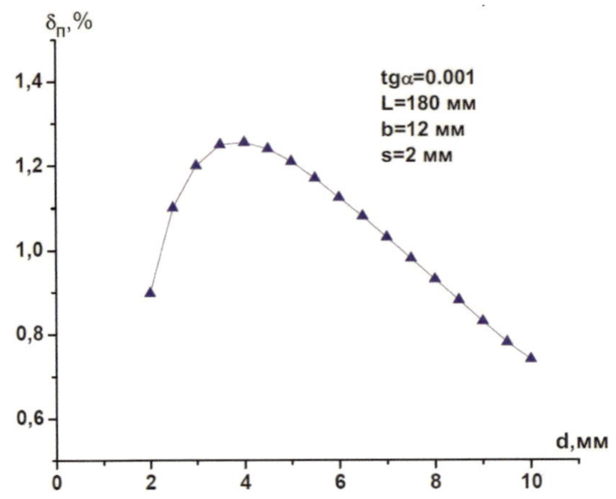

Figure 7.
Air gap as error for warp $\alpha \approx tg\ \alpha = 0.001$.

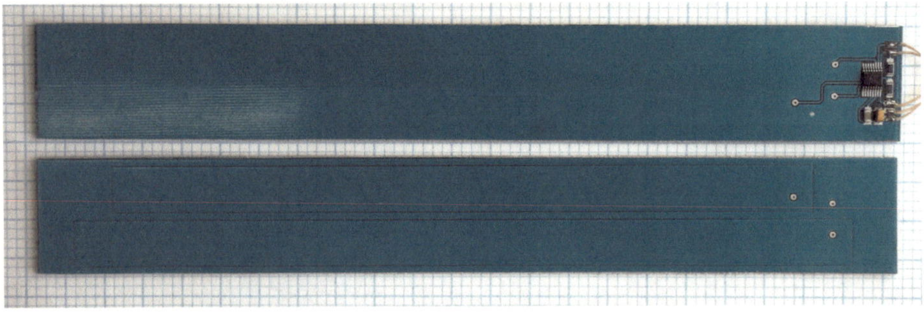

Figure 8.
Realized sensors in the department of electric and magnetic measurements of IED NASU: top – reverse side of the sensor; bottom – operational side of the sensor with working electrodes.

an initial setup or during a planned inspection of the hydrogenerators. In the operator the hydrogenerators have temperature nearly 70–80°C.

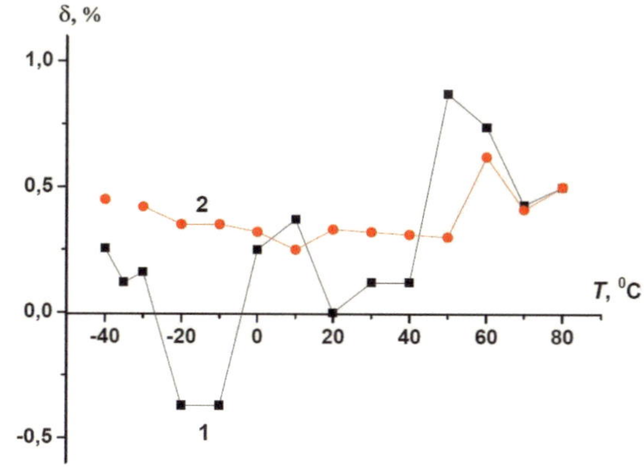

Figure 9.
A plot of the capacitance values in the temperature range from −30 to +80°C: 1 – the curve of capacitance C12P for the direct change; 2 – the curve of capacitance C12Z for the reverse change.

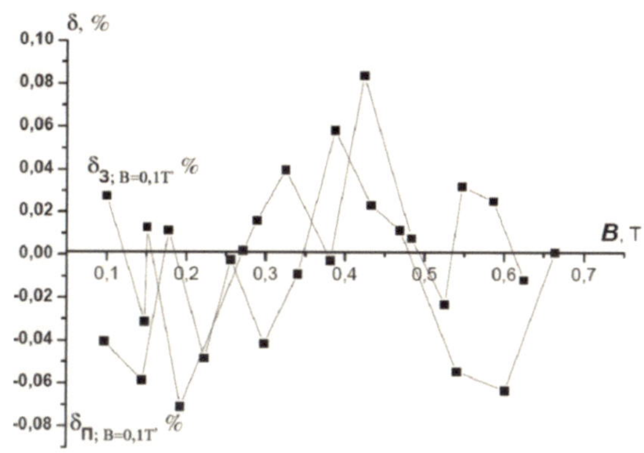

Figure 10.
The test sensing element for air gap HCS: $\delta_{\Pi;B=0,1\ T}$ for the direct change induction of electromagnetic field and $\delta_{3;B=0,1\ T}$ – for the reverse change induction of electromagnetic field.

Study influence of temperature stability is described detail in work [51]. And obtained result of the study's influence of the temperature on the stability of the air gap sensor was shown in **Figure 9**.

6.3.3 Electromagnetic stability

For experiments was designed special stand. The values of electromagnetic field industrial frequency are described in the works [52, 53]. The stand works as follows: variable voltage from the source is creating magnetic field in the air gap of coil. This magnetic field is modeling of electromagnetic field industrial frequency in turbo generator. In experiment values of electromagnetic field varied in range between 0.1 and 0.68 T. The principle of operation and the scheme of a special stand are described detail in following works [54].

The results of the natural experimental study of electromagnetic field influence on the measurement accuracy error of the air gap sensor shown in **Figure 10**.

Study industrial frequency magnetic field influence on microcontroller functioning stability is described as detail in work [55].

7. Conclusion

1. The structure and principle of operation of hybrid electro-optical sensors for the fault diagnosis system of large generators presented. Using fiber optic for data transmission allows you to easily and economically solve the implementation of anti-noise sensors for monitoring, control and measurement systems that are not subject to electromagnetic interference, electrical interference, explosive, can work in high-voltage or high-temperature environments. The hybrid sensor of the proposed design can be easily adapted to measure various parameters of mechanical defects in large generators.

2. It is shown that the skew of the sensor plane relative to the stator bore leads to technological errors.

3. The obtained experimental results confirm the possibility of using the proposed coplanar air gap sensor for air control systems in capsular hydrogenerators of the SGK 538/160-70M type.

The obtained results set the efficiency of the capacitive air gap sensor under the influence of industrial frequency magnetic field from 0.1 to 0.63 T.

Author details

Ievgen O. Zaitsev* and Anatolii Levytskyi
Department of Electric and Magnetic Measurements, The Institute of Electrodynamics of the NAS of Ukraine, Kyiv, Ukraine

*Address all correspondence to: zaitsev@i.ua

IntechOpen

© 2020 The Author(s). Licensee IntechOpen. Distributed under the terms of the Creative Commons Attribution - NonCommercial 4.0 License (https://creativecommons.org/licenses/by-nc/4.0/), which permits use, distribution and reproduction for non-commercial purposes, provided the original is properly cited. BY-NC

References

[1] Rassovsky VL. Ukrhydroenergo 15 years old. Hydroelectricity in Ukraine. 2018;**3-4**:4-7

[2] Chepulsky YP, Zamaraev KS. Vibration Diagnostics of Machine Parts. Moscow: MIIT; 2009. 75p

[3] Zaitsev IO. Opto-electronic transducer for large hydro generator shaft run-out fault diagnostic system: Design and measurement principle [thesis]. Kiev: IED NANU; 2019

[4] Levytskyi A, Fedorenko G, Gruboi O. Control of the State of Powerful Hydro and Turbogenerators by Means of Capacitive Measuring Instruments of Mechanical Defects Parameters. IED NANU: Kiev; 2011. 242p

[5] Babak SV, Myslovich MV, Sysak RM. Statistical Diagnostics of Electrical Equipment. IED NANU: Kiev; 2015. 456p

[6] Rolim F, Tetreault A, Marshall R. Air Gap Monitoring System Key Element to Correctly Diagnose Generator Problems. Belém city, Para state, Brazil: ENAM; 2004. 9p

[7] Date sheet: Using the VM600 System to Measure Air Gaps on Hydro Turbine. London: Zeefax Inc; 2012. 43p

[8] Zaitsev IO. Electric machines faults monitoring system with hybrid electro-optic capacitive mechanical sensors. In: Abstracts of the 2017 IEEE International Young Scientists Forum on Applied Physics and Engineering (YSF-2017); 17-20October 2017; Ukraine, Lviv: IEEE; 2017. pp. 15-18

[9] Zaitsev IO, Levytskyi AS, Kromplyas BA. Hybrid capacitive sensor for hydro- and turbo generator monitoring system. In: Proceedings of the International conference on modern electrical and energy system (MEES-17); 15–17 November 2017; Ukraine. Kremenchuk: IEEE; 2017. pp. 288-291

[10] Griščenko M. Air gap monitoring unit generator to advance vibration diagnostic procedure: Summary of the Doctoral Thesis. Riga: RTU; 2015. 20p

[11] Mikalauskas R, Volkovas V. Air gap modelling and control possibilities in rotary systems. Ultragarsas. 2003;**1**(46): 7-11

[12] Volkovas V, Gečys SJ. Increase the initial eccentricity of electric cars in experimental working conditions. Elektrotechnika : mokslo darbai. 1988; **15**(24):37-52

[13] Jonušas R, Jurkauskas A, Volkovas V. Rotorinių sistemų dinamika ir diagnostika. Kauno technologijos. 2001;**1**:213-228

[14] Date sheet: Air gap measuring system LS 120–121 transducers & ILS 730–731 conditioners. Vibro-Meter [Internet]. 2019. Available from: https://www.ahrinternational.com/PDF_catalogues/SKF/058406.pdf [Accessed: 30-06-2019]

[15] Manual: Iris Power Capacitive Air Gap Sensor. Catalog Qualitrol [Internet]. 2019. Available from: https://irispower.com/wp-content/uploads/2016/11/Iris-Power-CGS-Brochure-Capacitive-Gap-Sensor-1.pdf [Accessed: 30 June 2019]

[16] Date sheet: 4000 Series Air Gap Sensor System. Bently Nevada* Asset Condition Monitoring Specifications and Ordering Information [Internet]. 2019. Available from: https://www.industrial.ai/sites/g/files/cozyhq596/files/2019-06/4000_series_air_gap_sensor_system-167885f.pdf [Accessed: 30 June 2019]

[17] Air gap monitoring in generators with fiber based sensor [Internet]. 2017. Available from: http://fo4energy.hubersuhner.com/en/News/1868 [Accessed: 20 July 2017]

[18] Gorbova G, Gorbov M, Meyer M. Precise capacitance calculation of sensing elements capacitive sensors by method of direct field strength determination. In: Proceedings of the lEfct Sensor 2002; 12–14 June 2002; USA, Orlando; 2002. pp. 1239-1243

[19] Turan J, Ovseník L, Turán J. Optically powered fiber optic sensors. Acta Electrotechnica et Informatica. 2005;**5**:7

[20] Alekseev BA. Determination of the State (Diagnosis) of Large Hydrogenerators. Moscow: ENAS; 2002. 144p

[21] Potashnik SI, Fedorenko GM, Vaskovsky YN, et al. Problems of increasing the reliability of powerful hydrogenerators with instability of the air gap. Hydroelectricity in Ukraine. 2006;**3**:6-10

[22] Vaskovsky YN, Fedorenko GM, Voshchinsky KV, et al. Modeling of electromagnetic and thermal processes in the damper of the rotor of a hydrogenerator with an uneven air gap between the rotor and the stator. Hydroelectricity in Ukraine. 2006;**3**: 11-17

[23] Kuzmin VV, Vakulenko AN. Operating modes and the most common damage to hydro generators of hydroelectric power plants of the Dnieper cascade. The influence of the parameters of the air gap between the rotor and stator of the hydrogenerator on the reliability of its operation. Hydroelectricity in Ukraine. 2005;**2**: 22-30

[24] Pasko VE, Fedorenko GM. The introduction of continuous monitoring of hydrogenerators is a revolutionary step in improving their reliability and efficient operation. Hydroelectricity in Ukraine. 2005;**2**:37-40

[25] Petrov YV. Unevenness of the air gap of the hydrogenerator and its effect on the stator vibration. Electric Stations. 1974;**1**:25-28

[26] Fedorenko GM, Saratov VO, Chernish VV. Having inflated the parameters of the gap to the characteristics of the hydro generator. Hydroelectricity in Ukraine. 2005;**4**: 26-33

[27] Fedorenko GM, Vaskovsky YN, Saratov VA, et al. Modeling the electromagnetic load of the damper system of the rotor of a hydrogenerator in synchronous modes with an eccentricity of the air gap. Hydroelectricity in Ukraine. 2006;**1**: 25-32

[28] Azuare C, Millan A. Stator deformation of large hydrogenerators and its effects on the machines. In: Proceedings of the International Council on Large Electric Systems (CIGRE 2004); 29 August–3 September 2004; France, Paris; 2004. pp. 1-7

[29] Kuzmin VV, Vakulenko AN. Hydrogenerator poles damage by electrical erosion-pattern, cause, diagnostics and failure prevention. In: Proceedings of the International Council on Large Electric Systems (CIGRE 2006); 26 August–1 September 2006; France, Paris; 2006. pp. 1-9

[30] Netseevsky AB, Geller AB, Schetinin AD. Elimination of the asymmetry of the air gap in the hydrogenerator type SV1340/150-96. Electric Stations. 1975;**1**:45-49

[31] AD7745/46 24-Bit Capacitance-to-Digital Converter with Temperature Sensor [Internet]. 2019. Available from: https://www.analog.com/media/en/tech

nical-documentation/data-sheets/AD7745_7746.pdf [Accessed: 30 June 2019]

[32] Zaitsev IO, Levytskyi AS. Determination of response characteristic of capacitive coplanar air gap sensor. In: Proceedings of the 2017 IEEE Microwaves, Radar and Remote Sensing Symposium (MRRS-2017); 30 June–2 August 2017; Ukraine, Kiev: IEEE; 2017. pp. 85-88

[33] Sunmee P, Borton D, Kang M, Nurmikko A, Song Y. An implantable neural sensing microsystem with fiber-optic data transmission and power delivery. Sensors. 2013;**13**: 6014-6031

[34] Levytskyi A, Zaitsev I. Hybrid fiber-optic measuring tools for control and diagnostic parameters of hydrogenerators. Hydroelectricity in Ukraine. 2016;**3-4**:32-33

[35] International Measurement Confederation (IMEKO) [Internet]. 2009. Available from: www.imeko.org [Accessed: 30 June 2019]

[36] Zaitsev IO, Levytskyi AS, Kromplyas BA. Capacitive distance sensor with coplanar electrodes for large turbogenerator core clamping system. In: Proceedings of the 2019 IEEE 39th International Conference on Electronics and Nanotechnology (ELNANO); 16–18 April 2017; Ukraine, Kiev; 2017. pp. 644-647

[37] Zadvornov SA. Issledovaniye metodov postroyeniya gibridnykh volokonno-optikoskikh izmeritel'nykh sistem. Moscow: MPEI; 2009. 119p

[38] Rosolem JB, Floridia C, Sanz J. Optical system for hydrogenerator monitoring. In: Proceedings of the International Council for Power Electroenergetical Systems CIGRE 2010; France, Paris; 2010. pp. 1-8

[39] Riesen S, Schubert U, Bett AW. GaAs photovoltaic cells for laser power beaming at high power densities. In: Proceedings of the 17th European Photovoltaic Solar Energy Conference; 22–26 October 2001; Germany; pp. 182-185

[40] Peña R, Algora C, Anton I. GaAs multiple photovoltaic converters with an efficiency of 45% for monochromatic illumination. In: Proceedings of the 3rd World Conference on Photovoltaic Energy Conversion; 11–18 May 2003; Japan, Osaka: Arisumi Printing Inc. pp. 228-231

[41] Dakin J, Brown R. Handbook of Optoelectronics. Boca Raton: CRC Press; 2006. 1563p

[42] PV-Cells for Optical Power Transmission [Internet]. 2019. Available from: https://www.ise.fraunhofer.de/content/dam/ise/de/documents/infomaterial/brochures/photovoltaik/14e_ISE_Flyer_III-V_OpticalPowerTransmission.pdf [Accessed: 30-06-2019]

[43] Herb J. Commercialization of New Lattice-Matched Multi-Junction Solar Cells Based on Dilute Nitrides. Subcontract Report NREL/SR-5200-54721. San Jose, California: NREL; 2012. 28p

[44] Wang MR, Chen RT, Sonek GJ, Jannson T. Wavelength-division multiplexing and demultiplexing on locally sensitized single-mode polymer microstructure waveguides. Optics Letters. 1990;**15**(7):363-365

[45] Haid M, Armbruster C, Derix D, Schöner C, Helmers H. 5 W Optical Power Link with Generic Voltage Output and Modulated Data Signal. In: Proceedings of the 1st Optical Wireless and Fiber Power Transmission Conference (OWPT2019); 23–25 April 2019; Japan, Yokohama

[46] Zaitsev IO, Levytskyi AS, Sydorchuk VE. Air gap control system for hydrogenerators. Devices and Methods of Measurements. 2017;**8**(2): 122-130

[47] Manual: VMTM Air Gap. Capacitive Measuring Chains [Internet]. 2019. Available from: http://vibrosystm.com/wp-content/uploads/9628-25D1A-103.pdf [Accessed: 30 June 2019]

[48] Volkovas V, Mikalauskas R, Eidukeviciute M. Air gap measuring system for purpose of diagnostics and condition monitoring [Internet]. 2019. Available from: http://yadda.icm.edu.pl/yadda/element/bwmeta1.element.baztech-article-BAT3-0033-0039/c/httpwww_bg_utp_edu _plartdiagnostyka2004volkovas.pdf [Accessed: 30 June 2019]

[49] Manual: Air Gap Monitoring System [Internet]. 2019. Available from: https://www.giga-tech.it/media/prodPdf/Doc.%20AirGap_93_AyIM6_it.pdf [Accessed: 30 June 2019]

[50] Levytskyi A, Zaitsev E, Kromplyas B. Determination of the response characteristic of the capacitive sensor of the air gap in the hydrogenerator СГК 538/160-70М. The Institute of Electrodynamics of the National Academy of Sciences of Ukraine. 2017;**43**:134-137

[51] Zaitsev IO, Levytskyi AS, Kromplyas BA. Characteristic of capacitive sensor for the air gap control system in the hydrogenerator. In: Proceedings of the First Ukraine Conference on Electrical and Computer Engineering (UKRCON); 29 May–2 June 2017; Ukraine, Kyiv: IEEE; 2017. pp. 390-394

[52] Milykh VI, Vysochin AI. Principles of calculation of the magnetic field in the end of the turbo generator in various regimes of its work. Electrical Engineering. 2010;**3**:19-24

[53] Milykh VI, Dubinina ON. Numerical calculation of the magnetic field in the end zone of the turbogenerator under load conditions. Electrical Engineering and Electrical Mechanics. 2003;**1**:64-69

[54] Zaitsev I, Levytskyi A, Kromplyas B, Panchyk M, Bereznychenko V. Study influence industrial frequency magnetic field on capacitive pressing sensor for large turbogenerator core clamping system. In: Proceedings of the Ukraine Conference On Electrical And Computer Engineering (UKRCON); 2–6 July 2019; Ukraine, Lviv: IEEE; 2019

[55] Zaitsev I, Levytskyi A, Kromplyas B, Panchyk M, Bereznychenko V. Study industrial frequency magnetic field influence on STM32F051K8T6 microcontroller functioning stability. The Institute of Electrodynamics of the National Academy of Sciences of Ukraine. 2019;**52**:80-86

Chapter 3

Tidal Turbine Generators

Faisal Wani, Jianning Dong and Henk Polinder

Abstract

Recently, tidal stream turbines have become a preferable mode of harvesting tidal energy. The main issue for low utilization of tidal energy is the high levelized cost of energy (LCoE) from tidal stream turbines. A major reason for this is the high operation and maintenance costs for submerged installations. A possible way of minimizing the LCoE and improving the availability is to use a flooded (or a wetgap) generator rather than a conventional airgap generator. Inside flooded generators, the gap between the stator and rotor is filled with the seawater. This architecture has the potential to improve cooling and reduce reliance on ancillary systems (e.g., bilge system), thereby improving reliability. The chapter begins with a brief description of the generator systems used in current tidal stream turbines. The focus of the chapter is, however, to give a basic insight into the design aspects of the flooded generators, and compares it with the currently used sealed airgap generators in tidal turbine systems.

Keywords: tidal stream turbines, permanent magnet generators, flooded generators, corrosion, rotor-can

1. Introduction

Energy from tides is predictable over a span of several years, unlike other renewable sources of energy such as solar and wind. Furthermore, tidal energy is influenced little by weather conditions. Grid operators prefer a predictable energy resource as it facilitates economic and reliable grid operation. According to [1], the estimated global potential for tidal energy is around 500 GW, as shown in **Figure 1**. There are two main ways of harnessing tidal energy: tidal range (or dams) and tidal stream turbines.

Recently tidal stream turbines have become a preferable mode of harvesting tidal energy over tidal range (or dams) technology. The commercial success of wind turbines is to a large extent responsible for this shift in harnessing tidal energy. Furthermore, the potential for tidal stream turbines is expected to be more than the tidal range technology [2].

A typical tidal stream turbine is shown in **Figure 2**, which is an example of a horizontal axis tidal turbine (HATT). Although vertical axis tidal turbines and other topologies such as oscillating hydrofoil, enclosed tip turbine and tidal kites are also used to harness tidal energy, most of the research and development efforts are focused on HATTs; see **Figure 3**. This is primarily because of their higher technology readiness level (TRL), and similarity to commercial wind turbines. Some of the main tidal energy developers are listed in **Table 1**. As is evident, most of them prefer horizontal axis tidal turbines. Consequently, the focus of this chapter will be also on HATTs.

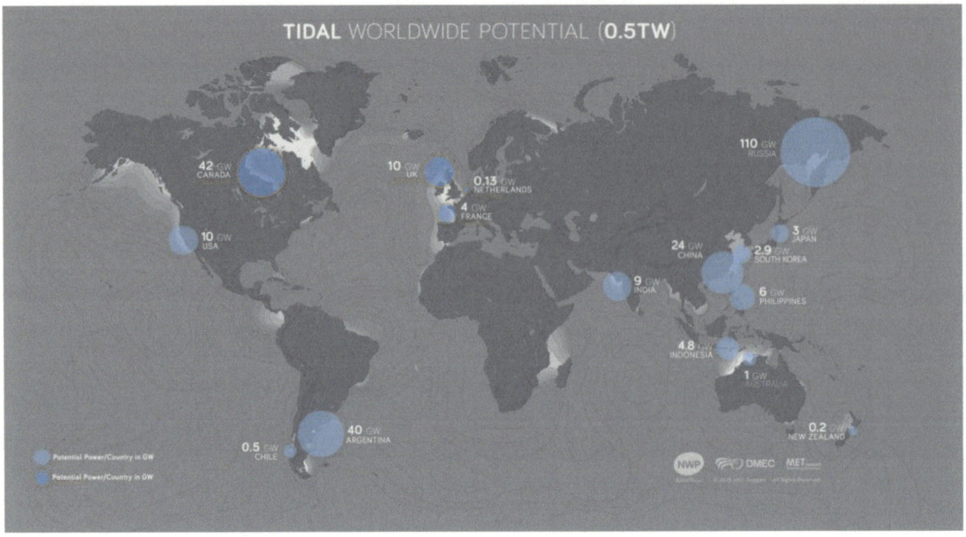

Figure 1.
Global tidal energy potential (Source [1]).

Figure 2.
Nova M100 tidal turbine (Source: ©Nova Innovation).

The power captured by a HATT is given by the same equation as that of a wind turbine. Mathematically, power captured by the turbine is given by

$$P = \frac{1}{2} C_p(\lambda) \rho A v^3, \tag{1}$$

where C_p is the power coefficient, ρ is the density of the water, A is the area swept by the turbine blades and v is the velocity of the tidal stream. The power

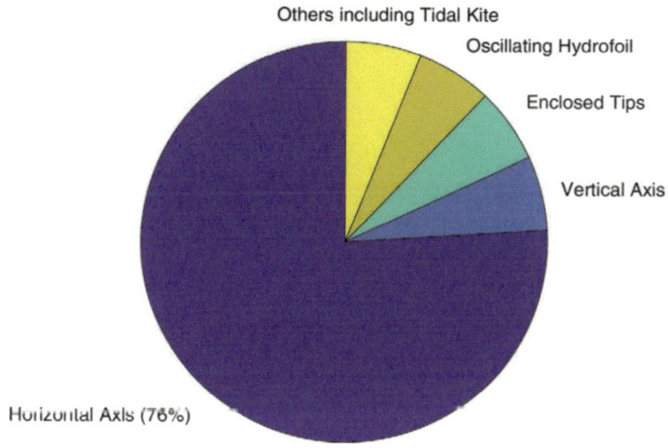

Figure 3.
Focus of R&D efforts for different tidal stream technologies (Source: [3]).

Company name	Country base	Device type	Generator type[a]	Device name[b]
Andritz Hydro Hammerfest	Norway	HATT	IG + GB	HS1000
Atlantis Resources Limited	UK	HATT	PMSG + GB	AR1500
Marine Current Turbines	UK	HATT	IG + GB	SeaGen S
Nautricity	UK	HATT	PMSG-DD	CoRMaT
Nova Innovation	UK	HATT	IG + GB	Nova M100
Schottel Group	Germany	HATT	IG + GB	SIT
Scotrenewables	UK	HATT	IG + GB	SR2000
Tocardo Tidal Turbines	Netherlands	HATT	PMSG-DD	T200
Minesto	Sweden	Tidal Kite	—	Deep Green
Deepwater Energy BV	Netherlands	VATT	—	Oryon Watermill

Source: European Marine Energy Centre (emec.org.uk).
[a]*IG: induction generator; GB: gearbox; PMSG: permanent magnet synchronous generator; DD: direct drive.*
[b]*Not all the devices from the same manufacturer are listed.*

Table 1.
Some of the main tidal energy developers around the world, and their device types.

coefficient, C_p is a function of the tip-speed ratio denoted by λ (for a constant blade pitch angle). Tip-speed ratio is the ratio of the blade tip-speed to the incoming velocity of the fluid. A typical $C_p - \lambda$ curve is shown in **Figure 4**.

A disadvantage of the tidal energy is that the levelized cost of energy (LCoE) is currently much higher than the average LCoE from more developed sources of energy [3]. In [3], minimizing the operation and maintenance (O&M) expenses, and increasing the capacity factor have been identified as major factors in minimizing the LCoE from tidal turbines; see **Figure 5**.

One possible way of minimizing the O&M expenses and improving the capacity factor is to use a flooded (or wetgap) generator rather than the sealed airgap generator [3]. In a flooded generator, instead of an airgap the stator-rotor gap is filled with the seawater. As a result, the design of a flooded generator has to be different from the airgap generator. To begin with in a flooded generator a stator and a rotor *can* (or sleeve) are required to protect the active machine parts from corrosion. This will be illustrated later in the chapter. Compared to the conventional airgap generators, the design of flooded generators has been little addressed in literature. We discuss some design aspects of the flooded generators in this chapter.

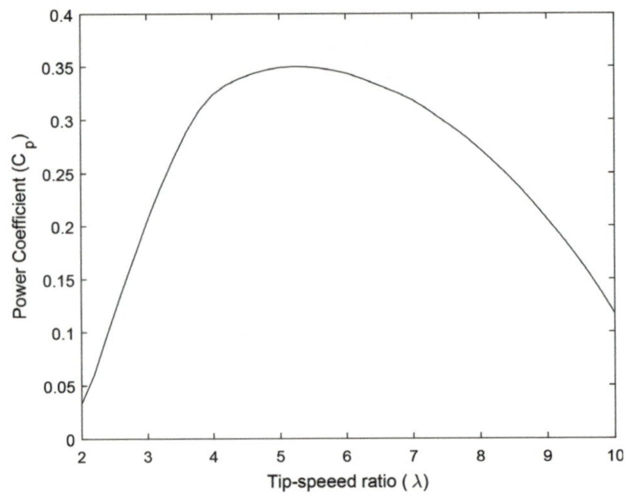

Figure 4.
A typical $Cp - \lambda$ for a tidal turbine [4]. Only indicative values are shown here; these values do not reflect the maximum Cp in a tidal turbine.

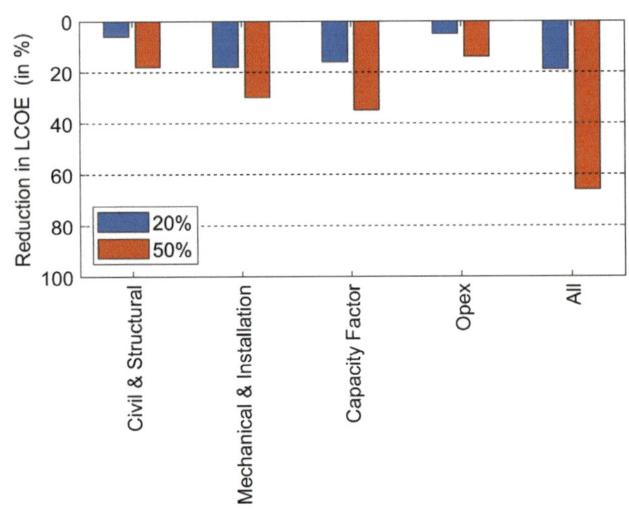

Figure 5.
Likely reduction in per unit LCoE for 20 and 50% reduction in cost component (Source: [3]).

This chapter gives a qualitative comparison between a conventional airgap and a flooded generator for tidal stream turbines. The qualitative comparison is based on the pros and cons of each generator type, followed by their structural differences. Both of these generators are assumed to be permanent magnet (PM) radial-flux direct-drive generators, unless otherwise stated. This design is used for comparison as the radial-flux PM machines are well suited to both flooded and conventional architectures.

In Section 2, a brief overview of generator technologies used in the current tidal stream turbines is given. Section 3 illustrates the conventional design of the generators used in tidal turbines, and their structural aspects and limitations. In Section 4, a relatively new concept called the flooded generator is proposed for application in tidal turbines. Section 5 gives a general guideline for design of flooded generator. Section 6 gives the conclusions from this chapter.

2. Generator types in tidal turbines

Squirrel cage induction machines and PM synchronous machines are commonly used as generators in tidal turbines. Among synchronous generators, PM machines are more attractive compared to the electrically excited synchronous generators, despite being expensive. This is primarily because of the lower failure rates and possibility of higher pole numbers in PM machines. Usually generators can be classified in either of the two categories: high-speed generators with a gearbox, and low-speed direct-drive generators [5].

2.1 High-speed generators with a gearbox

The size of the generator is directly related to its torque rating. This implies for the same power, high-speed machines are smaller. In addition to the manufacturing costs, transport and assembly costs are also likely to be lower for high-speed machines. Since tidal turbines rotate at low speeds, a gearbox is necessary between the turbine and the high-speed generator; see **Figure 6a**.

Atlantis Resources AR1500 turbine uses a radial flux surface-mounted PM synchronous generator with a two-stage epicyclical gearbox [6]. On the other hand, Andritz Hydro Hammerfest HS1000 and Schottel's Instream turbine comprises an induction generator with a three-stage and a two-stage planetary gearboxes, respectively [7, 8].

2.2 Low-speed direct-drive generators

Low-speed direct-drive generators can be directly coupled to the turbine without the need for a gearbox; see **Figure 6b**. This significantly improves the overall

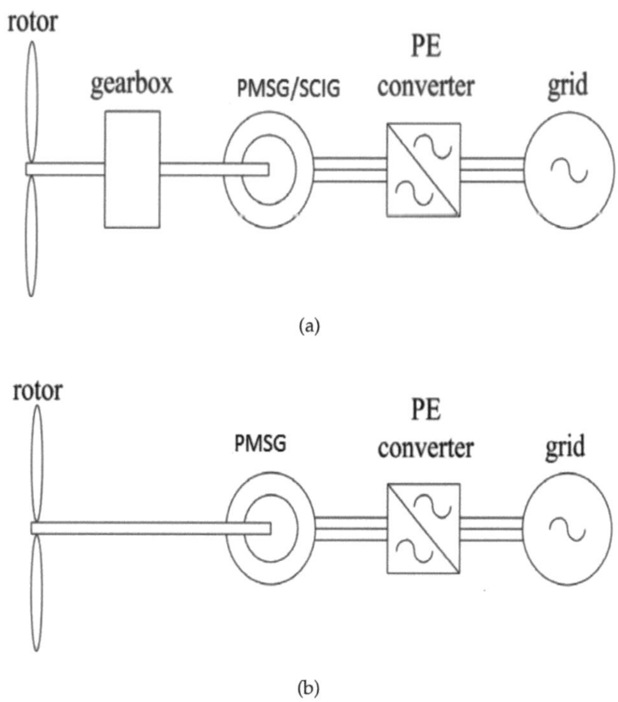

Figure 6.
(a) High-speed generator coupled to the tidal turbine via a gearbox; (b) low-speed direct drive generator. PMSG—Permanent magnet synchronous generator; SCIG—Squirrel cage induction generator; PE—Power Electronic.

reliability and efficiency of the system. Direct-drive generators usually have high pole numbers, and thus small pole-pitch. This makes PM machines suitable for direct-drive applications because induction machines with high pole numbers result in an inefficient design due to lower power factor [9].

DCNS-OpenHydro and Tocardo T2 turbines respectively, used a direct-drive PM generator. Nautricitys CoRMaT is another innovative direct-drive generator with two contra-rotating rotors. CoRMaT's design is unconventional as the stator of the axial-flux PM generator is also mounted on one of the rotors to double the relative speed between the stator and the rotor. The torque is equally divided between the two rotors. Because of its peculiar design, the device is always perpendicular to the tidal flow ensuring maximum capture of the power [10, 11].

3. Conventional design for tidal generators

Most tidal turbine manufacturers use a conventional airgap generator. In an airgap generator, the stator-rotor gap inside the generator is occupied by air. Such generators are almost invariably used in energy generation, such as hydro power stations, and wind turbines. For tidal turbines, the airgap generator has to be enclosed in a water-tight nacelle. The inner nacelle space is isolated by the seawater using a high-pressure rotary seal on the rotating shaft of the turbine, as shown in **Figure 7**. Sealing the nacelle protects against the corrosion and/or electrical breakdown of critical components such as the generator coils and magnets, and bearings.

The drawback with this topology is that it makes the power take-off (PTO) system vulnerable to any water ingression in the nacelle, mostly via the leakage through the seals [3, 12]. Water-tight rotary seals on the shaft are also susceptible to wear, and generate losses due to friction. Besides decreasing efficiency, these losses may end up heating the bearings and compromising their lifetime [12]. To avoid high friction losses, small leakage of the fluid is allowed, which acts as a lubrication film between the seal faces. Whereas, a thicker fluid film decreases wear, it will also

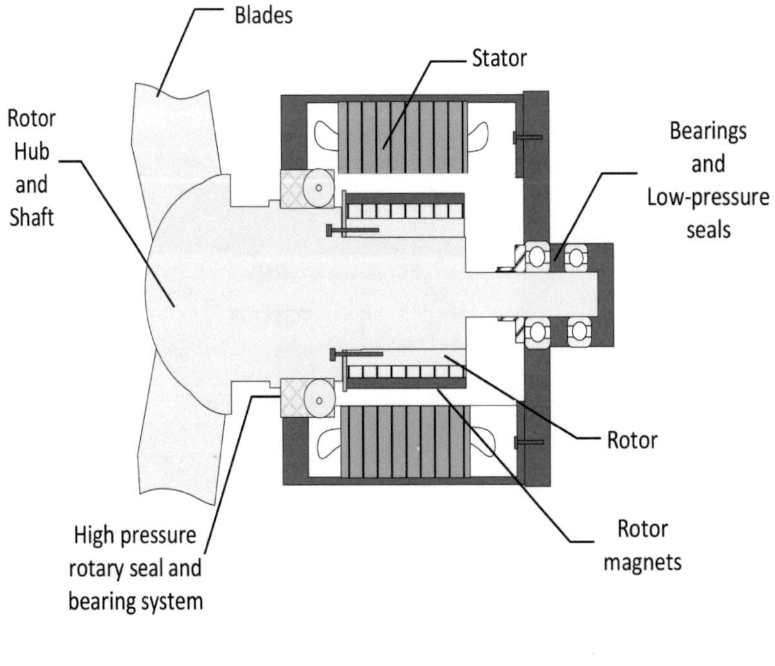

Figure 7.
A sealed airgap direct-drive generator; only necessary details for discussion are shown here.

increase leakage. Normally in ships or submarines this leakage water accumulated over time is drained using a bilge pump. Having a bilge pump in a tidal turbine will add to the cost, and may compromise reliability. However, large tidal turbines of the order of 1 MW do use bilge pumping systems alongside dehumidifiers. In addition to wear, seals usually fail when subjected to unforeseen or abnormal operating conditions [13]. This includes foreign material such as sediments intrusion in the sealing faces, imperfect shaft alignment, corrosion, sticking and clogging.

From above it is clear that the seal breakdown or deterioration is a likely possibility in tidal turbines, and thus regular maintenance of the seals is necessary. Even performing preventive maintenance can increase the O&M expenditure if the maintenance cycle is frequent or poorly timed. LCoE studies for tidal energy systems suggest that for competing with other renewable energy sources, tidal turbines must survive for about 20 years, with a service interval of 5 years [14]. Unfortunately, not much data is publicly available about the reliability of seals in tidal turbines, still it is possible to have a reasonable estimate about seal reliability from other applications. For instance, the failure rate of seals in marine propellers [15], or canned motor pumps [16], can provide a rough guide about the seal failure rates in tidal turbines. From the failure rate of seals in propellers and canned motor pumps, it appears challenging to completely rely on high-pressure seals for the aforementioned maintenance interval without any failure or leakage. This has led to a discussion about the possible use of flooded generators in tidal turbines [14, 17].

Despite relying on the seals for their proper functioning, airgap generators have their own advantages. First of all, it is a time-tested architecture and rules-of-thumb often suffice to design a generator with desired requirements. Secondly, because of no additional sleeves in the airgap, the effective airgap length is smaller. This means that the required magnet volume is less, reducing the cost of generator. Moreover, no sleeves also mean manufacturing process is simpler and less expensive. Thirdly, if the sleeve material is electrically conductive, additional eddy current losses in the machine will reduce the efficiency of the generator. Lastly, the drag losses in the airgap are lower than in the watergap.

4. Flooded generators

Among early tidal turbine developers, DCNS-Openhydro and Voith tried to use this topology. Nautricity's CORMAT and NOVA Innovation's TiPA are two examples of tidal turbines where flooded generators are used. Envisaging the use of flooded generators in tidal energy generation has been a recent development. On the other hand, their motor counterparts—commonly known as canned motors—have been around for quite some time [18].

Inside a flooded generator, not only electromagnetic parts but also other components such as bearings are susceptible to corrosion. Same can be said about the canned motors. As far as bearings are concerned, as a starting point it might be reasonable to consider the bearings used in canned motors. Normally canned motors are used in pump applications [16]. This means there is usually a high-pressure fluid available which can be used as a lubricant in either a hydrostatic or a hydrodynamic bearing. It is obvious that non-corrosive materials with good resistance to abrasion must be used in the bearings.

In an ideal scenario, flooded generators would also be equipped with seawater-lubricated hydrodynamic bearings. But there are obvious problems to this solution for tidal turbines. First, the tidal turbines usually run at low speeds (<40 rpm); and second, the seawater has very low viscosity. Both of these would mean too thin a

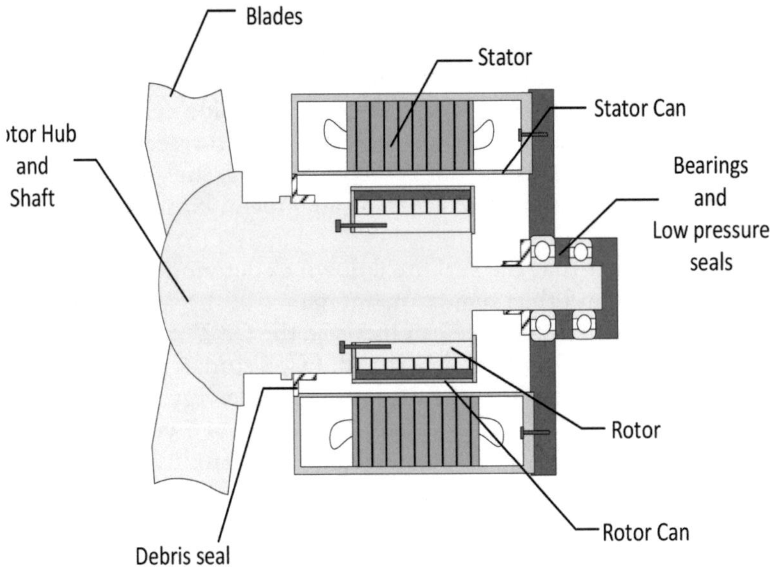

Figure 8.
A representative figure of a flooded generator; only necessary details for discussion are shown here.

fluid film thickness for good hydrodynamic lubrication [12]. Hydrostatic water-lubricated bearings would require an external motor-driven pump, which not only adds to the cost but also creates another potential fault point. Therefore, most tidal turbine manufacturers prefer commonly used roller bearings with some oil/grease lubrication.

The drawback with using roller bearings is that they are susceptible to corrosion, silt penetration and fatigue due to cyclic stresses. In such a case, a high pressure seal is still required, but it is better protected from debris and marine growth, and can often be smaller than in the sealed airgap design. Also, in a flooded generator, it is also possible to fill the space inside the generator with fresh water at a slightly higher pressure than the ambient pressure. Filling the generator with clean/fresh water might increase the life of any hydrodynamic bearings inside the generator.

To avoid fouling or clogging of organic material in the flooded gap, it is likely that a debris seal will be used in tidal turbines, thereby preventing the free water circulation in the watergap [19]. These seals are in addition to the bearing seals, as illustrated in **Figure 8**. However, debris seals are not required to provide high-pressure sealing—but only block large particles including marine life from entering the generator—which means they are cheaper and require less maintenance.

The safeguarding of the electromagnetic parts of the generator—the stator windings and the rotor magnets—is addressed in the next section.

5. Design choices for flooded generators

It has already been stated that the main difference between the airgap and the flooded generators is the presence of stator and rotor *cans* in the latter. These *cans* are used to enclose the stator windings and rotor magnets, and prevent their exposure to the seawater. Previous sections mainly dealt with the structural features of the conventional and flooded generators. The focus of this section is to look at the flooded generator design more from the electromagnetic and thermal perspectives.

DOI: http://dx.doi.org/10.5772/intechopen.90433

5.1 Rotor-can materials

In **Figure 8** we see that a protective layer of some material has to be used to protect the active electromagnetic parts of the generator against contact with water. In literature, we find instances of both metallic as well as non-metallic materials being used for the rotor-can in flooded generators or canned motors [20].

With metallic materials, such as stainless steel, a high eddy-current loss in the rotor can or may occur because of the asynchronously rotating components of the stator magneto-motive forces (MMF). However, the magnitude of these losses could either be significant or insignificant depending on the winding layout in the stator, wetgap dimensions, thickness of stator and rotor-cans, conductivity of the material and electrical current loading in the stator. A PM machine with integer slot distributed winding usually results in very low losses for low-speed direct drive generators, even with conductive rotor-can material [21]. **Figure 9** shows the eddy current loss density in the rotor-can and the permanent magnets. The image is taken from a study conducted on comparing different rotor-can materials in a flooded generator [20]. The PM generator in the study is rated at 300 kW at 30 rpm. The winding layout used in the machine is a fractional slot winding (total no. of slots = 132, no. of poles = 120), and thus has a relatively high loss in the rotor. The total eddy-current loss in the rotor from the stator MMF is about 5 kW, which is about 1.7% of the rated power.

Obviously, for non-conductive can material such as Carbon/Glass fiber, eddy-current loss in the rotor-can is negligible. However, the catch with using fiber materials is that water-proofing needs a thicker can. This increases the total magnetic reluctance, which implies thicker magnets are required to set up adequate magnetic field increasing the cost of the generator. Moreover, a relatively poor thermal conductivity may result in slight increase in the rotor temperature [20].

In a more novel concept of using a flooding generator with active magnetic bearings, a non-conductive material would be more suited for the rotor-can. A conductive rotor-can would result in very high losses, and could degrade the performance of the magnetic bearings because of the shielding effect of the rotor-can.

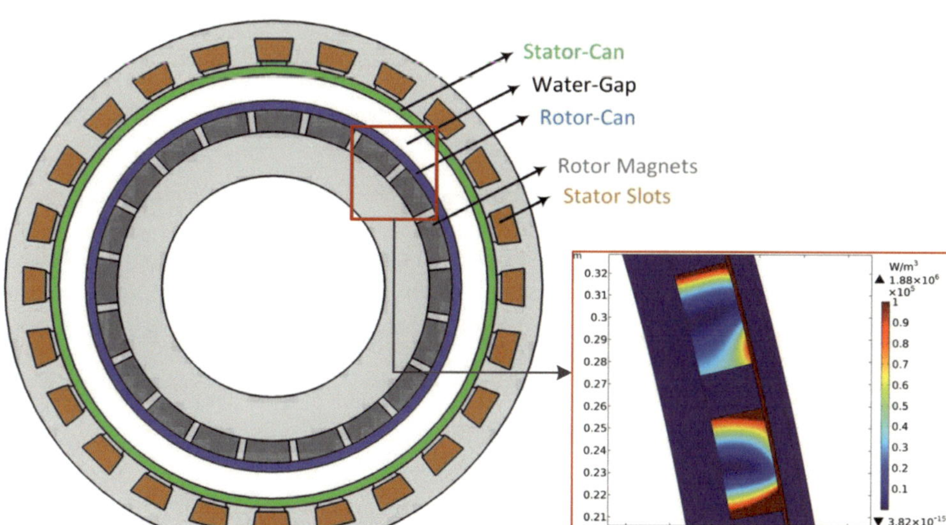

Figure 9.
Eddy current loss density in rotor-can and magnets of a test case PM flooded generator. For sake of clarity, the cross-section of the generator shown in the figure does not show all the stator slots and magnet poles of the test case PM generator.

However, the magnitude of losses and the shielding effect is a function of the speed. Hence, such effects could be less critical in low-speed tidal generators. A detailed analysis should thus be carried out before drawing any sweeping conclusions.

In another study, it was also found that the losses induced in the rotor from higher order time harmonics introduced from the switching of the power converter are insignificant for low-speed generators [21]. **Figure 10** shows the contributions from different frequency components of the stator current for the same 300 kW PM machine as above.

5.2 Stator-can materials

In principle, it is possible to use the same materials for the stator-can as the rotor-can. However, the use of metallic materials for the stator-can will result in excessive eddy current losses [20]. This does not mean that designs with metallic

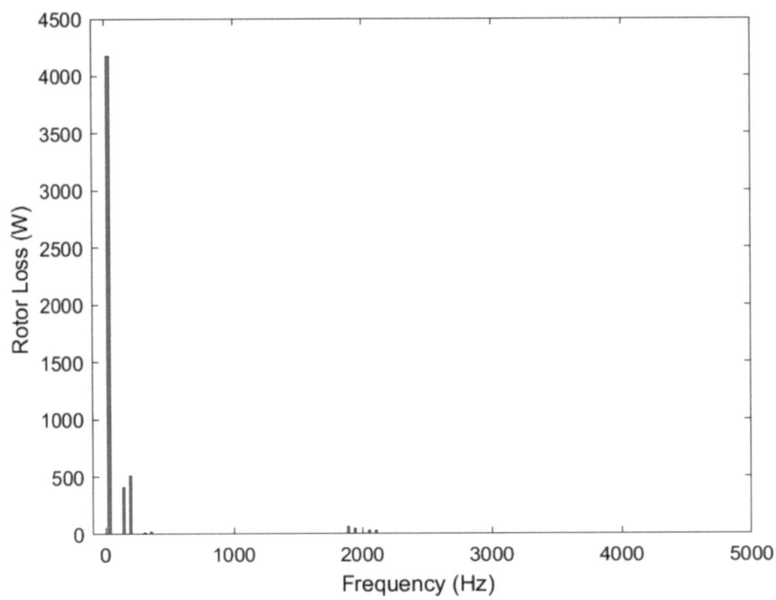

Figure 10.
Rotor eddy-current loss in the stainless steel rotor-can as a function of time harmonic component of current. Switching frequency is 2 kHz.

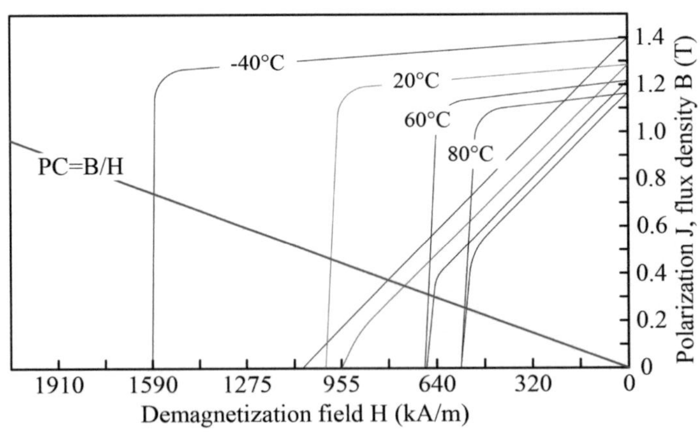

Figure 11.
A typical BH-curve for a NdFeB type permanent magnet.

stator-cans are never used. In fact canned motors have in the past invariably used metallic materials on the stator-can, and in some applications, materials such as 316 Stainless steel or Hastealloy 276, still continue to be used as stator-cans [16]. However, these days, materials such as carbon fiber or glass fiber can offer adequate protection against water ingression, albeit again making the stator-can thicker compared to a metallic-can. This increased thickness of the stator-can is more justified than the corresponding increase in the rotor-can thickness because of the significant improvement in efficiency. Obviously, such design choices also need to consider the ease in manufacturing.

As mentioned earlier, increased thickness of the can material increases the total magnetic airgap, thereby increasing the magnet volume and total generator cost. However, improved efficiency can recover the initial cost. The final choice must be carefully made based on whether a good protection can be provided by non-metallic

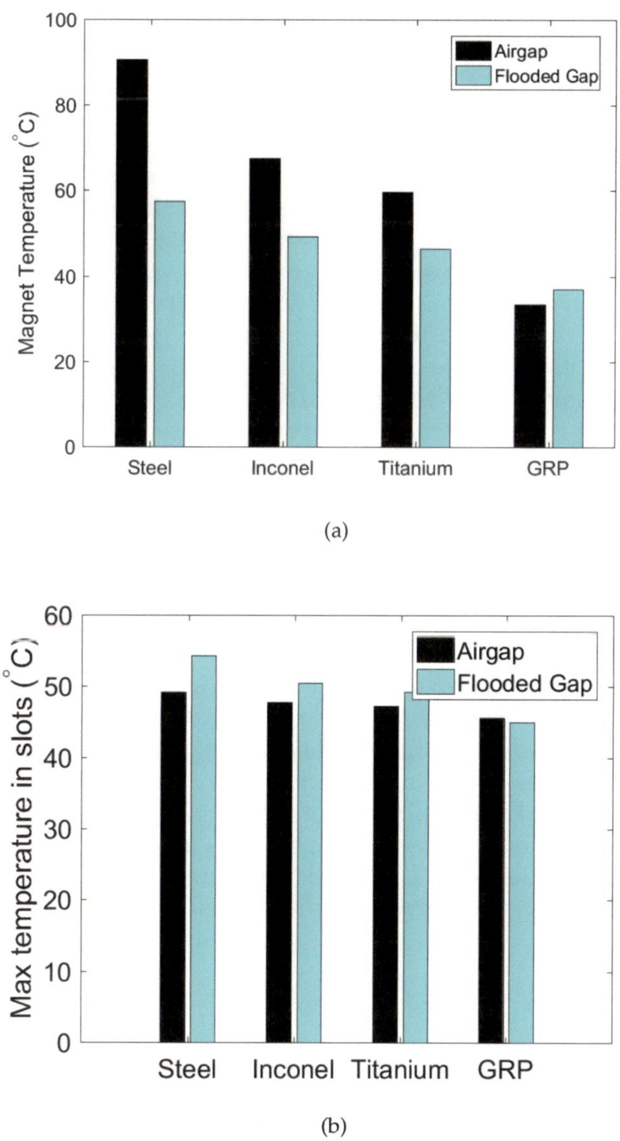

Figure 12.
(a) Temperature inside permanent magnet for different rotor-can materials; (b) maximum temperature in the stator slots for different rotor can materials. Cooling effect of flooded gap is clearly evident [20].

materials, and its structurally feasibility. If yes, then they should be selected over metallic materials.

5.3 Thermal considerations

Another point in the favor of flooded generators is the possibility of better cooling. This means flooded generators could be designed with high power density. For permanent magnet generators, this can happen in a two-fold manner. Firstly, the current density in the stator can be higher. And secondly, the magnets can operate at a higher permeance coefficient and the risk of demagnetization is lower. This can be seen from the BH-curve of a typical NdFeB permanent magnet, as illustrated in **Figure 11**. In other words, the temperature distribution within the machine is more uniform. However, this should not be universally assumed to be true. As Judendorfer et al. [22] suggest, it is possible that the significant improvement in cooling may only be achieved if forced circulation of water is possible.

The presence of water in the stator-rotor gap of the flooded generator facilitates the transfer of heat between the stator and the rotor. This happens even if the water in the gap is not continuously replenished with the ambient seawater. Whether the flow of heat is from the rotor to the stator or vice-versa depends on the stator losses, rotor losses (including eddy current losses in the can and permanent magnets), thermal impedance from stator to ambient and from rotor to ambient via bearing, shaft, etc.

As illustrated in **Figure 12**, the difference in the stator slot and the rotor magnet temperature is lower in the flooded generator as compared with the airgap generator (also with a similar rotor-can). **Figure 12** is taken from [20], where an airgap generator is compared with a flooded generator; both generators are assumed to have rotor-cans to illustrate the transfer of heat in the watergap. Note that the presence of water in the stator-rotor gap may result in either the increase of average temperature of the stator or rotor. This is provided that all the external thermal impedances remain the same.

6. Conclusions

In this chapter a brief overview of generators used in tidal turbine systems was given. The drawbacks of the conventional generator designs for tidal turbine applications were discussed. Although conventional designs are normally designed to operate for longer terms and there is a sense of reliability in the design, innovative solutions for submerged applications will be required. The primary consideration (even more than the efficiency) must be reliability. Lower failure rates of the generator would mean fewer maintenance expenses, and lower cost of energy. Flooded generators were proposed for applications in the tidal stream turbines. Different design considerations for flooded generators were explained in this chapter to give reader an overview about the design guidelines of such generators.

Acknowledgements

The authors have been supported by the TiPA (Tidal turbine Power take-off Accelerator) project, which has received funding from the European Union's Horizon 2020 research and innovation program under Grant Agreement No. 727793, managed by the Innovation and Networks Executive Agency. This chapter reflects only the authors' views, the Agency is not responsible for any use that may be made of the information the chapter contains.

Author details

Faisal Wani[1]*, Jianning Dong[2] and Henk Polinder[1]

1 Department of Maritime and Transport Technology, Delft University of Technology, Delft, The Netherlands

2 Department of Electrical Sustainable Energy, Delft University of Technology, Delft, The Netherlands

*Address all correspondence to: f.m.wani@tudelft.nl

IntechOpen

© 2020 The Author(s). Licensee IntechOpen. Distributed under the terms of the Creative Commons Attribution - NonCommercial 4.0 License (https://creativecommons.org/licenses/by-nc/4.0/), which permits use, distribution and reproduction for non-commercial purposes, provided the original is properly cited.

References

[1] Focal Point for Marine Energy [Online]. Available from: https://www.delta.tudelft.nl/article/focal-point-marine-energy [Accessed: 03 November 2019]

[2] Kempener R, Neumann F. Tidal Energy Technology Brief [Online]. Available from: https://www.irena.org/documentdownloads/publications/tidal_energy_v4_web.pdf [Accessed: 03 November 2019]

[3] Magagna D, Monfardini R, Uihlein A. JRC Ocean Energy Status Report 2016 Edition. Luxembourg: Publications Office of the European Union; 2016. Available from: http://www.orizzontenergia.it/download/Appr/ENERGIA%20DAL%20MARE/2017_03_31_Ocean_Energy_Status_Report.pdf [Accessed: 08 September 2019]

[4] Bahaj AS, Molland AF, Chaplin JR, Batten WMJ. Power and thrust measurements of marine current turbines under various hydrodynamic flow conditions in a cavitation tunnel and a towing tank. Renewable Energy. 2007;**32**(3):407-426

[5] Wani F, Polinder H. A review of tidal current turbine technology: Present and future. In: 2017 XIIth European Wave and Tidal Energy Conference (EWTEC); 2017

[6] Atlantis Resources Limited. AR1500 Tidal Turbine [Online]. 2016. Available from: https://www.atlantisresourcesltd.com/wp/wpcontent/uploads/2016/08/AR1500-Brochure-Final-1.pdf [Accessed: 03 July 2019]

[7] Schottel Hydro. SIT Instream Turbine [Online]. 2010. Available from: https://www.schottel.de/fr/schottel-hydro/sit-instream-turbine/ [Accessed: 03 July 2019]

[8] Andritz Hydro Hammerfest. Tidal Turbines [Online]. 2017. Available from: http://www.andritzhydrohammerfest.co.uk/tidal-turbines/ [Accessed: 03 August 2019]

[9] Dubois MR, Polinder H, Ferreira JA. Comparison of generator topologies for direct-drive wind turbines. In: Proceedings of the 2000 NORPIE; 2000. pp. 22-26

[10] NAUTRICITY. How CoRMaT Works [Online]. 2014. Available from: http://www.nautricity.com/cormat/cormat-efficiency/ [Accessed: 23 May 2019]

[11] NAUTRICITY. CoRMaTs Advantage [Online]. 2014. Available from: http://www.nautricity.com/cormat/cormat-power/ [Accessed: 23 May 2019]

[12] Harnoy A. Bearing Design in Machinery: Engineering Tribology and Lubrication. Basel, Switzerland: Marcel Dekker, Inc.; 2002

[13] Towsyfyan H, Gu F, Ball A. A Review of Mechanical Seals Tribology and Condition Monitoring [Online]. University of Huddersfield. Available from: http://eprints.hud.ac.uk/id/eprint/19391/2/towsyfyan.pdf [Accessed: 05 September 2019]

[14] Djebarri S, Charpentier JF, Scuiller F, Benbouzid M, Guemard S. Rough design of a double-stator axial flux permanent magnet generator for a rim-driven marine current turbine. In: 2012 IEEE International Symposium on Industrial Electronics. IEEE; 2012. pp. 1450-1455

[15] Carlton JS. Podded propulsors: Some results of recent research and full scale experience. Journal of Marine

Engineering & Technology. 2008;7(1): 1-4

[16] Scott C. Considerations for a canned motor pump. World Pumps. 2004; **2004**(459):22-25. Available from: https://www.sciencedirect.com/science/article/pii/S0262176204004195 [Accessed: 23 May 2019]

[17] Drouen L, Charpentier JF, Semail E, Clenet S. Study of an innovative electrical machine fitted to marine current turbines. In: OCEANS 2007-Europe. IEEE; 2007. pp. 1-6

[18] Lipe BR, Webb CR, inventors; Allis-Chalmers Corp, assignee. Canned motor. United States Patent US 2,958,292. 1960

[19] Holstein B, Perner N, inventors; Voith Patent GmbH, assignee. Submersible power generating plant, driven by a water flow. United States Patent US 8,410,626. 2013

[20] Wani F, Polinder H, Dong J, Yadav A. Comparing different materials for rotor-can in flooded generators. In: 2018 XIII International Conference on Electrical Machines (ICEM). IEEE; 2018. pp. 2572-2578

[21] Wani F, Shipurkar U, Dong J, Polinder H. Calculation of PWM-induced rotor-can losses in flooded generators. In: 2019 XIIIth European Wave and Tidal Energy Conference (EWTEC); 2019

[22] Judendorfer T, Fletcher J, Hassanain N, Mueller M, Muhr M. Challenges to machine windings used in electrical generators in wave and tidal power plants. In: 2009 IEEE Conference on Electrical Insulation and Dielectric Phenomena. IEEE; 2009. pp. 238-241

Section 2

Wind Generators

Chapter 4

Ocean Wind Energy Technologies in Modern Electric Networks: Opportunity and Challenges

Foad H. Gandoman, Abdollah Ahmadi, Shady H.E. Abdel Aleem, Masoud Ardeshiri, Ali Esmaeel Nezhad, Joeri Van Mierlo and Maitane Berecibar

Abstract

Wind energy is one of the most important sources of energy in the world. In recent decades, wind as one of the massive marine energy resources in the ocean to produce electricity has been used. This chapter introduces a comprehensive overview of the efficient ocean wind energy technologies, and the global wind energies in both offshore and onshore sides are discussed. Also, the classification of global ocean wind energy resources is presented. Moreover, different components of a wind farm offshore as well as the technologies used in them are investigated. Possible layouts regarding the foundation of an offshore wind turbine, floating offshore, as well as the operation of wind farms in the shallow and deep location of the ocean are studied. Finally, the offshore wind power plant challenges are described.

Keywords: ocean wind energy, offshore wind energy conversion, offshore renewable energy, power transmission, offshore wind turbines, active and reactive powers

1. Ocean wind energy resource

The increasing demand for electric power, the limited availability of fossil fuels, and increased environmental pollution have made it essential to Integra Clean Energy sources such as wind in our energy systems. On the other hand, the shortage of drought and varieties of geographical possibilities justifies the approach to developing offshore wind farms.

The offshore wind farms are wind turbines that are built several kilometers offshore in the ocean or sea for more efficient utilization of wind energy. Although this method is already very costly, increasing technological advances in turbines materials and bases, composite structures, as well as the construction of multimegawatt generators accelerate the deployment of offshore wind farms and make them a huge part of future energy production [1–3].

This chapter starts with introduction to the wind energy resource. After that, the global ocean wind energy resource is presented. Ocean wind energy technologies

IntechOpen

are explained in the next section. Then, the possible structure of offshore wind turbine is considered. Finally, challenges of offshore wind power are discoursed.

One of the renewable energy resources that can be used to generate electricity is ocean wind. Two-thirds of the Earth's surface is water; this potential could be used to generate electricity in different parts of the planet. However, using the ocean wind source varies depending on the geographical conditions and seasons. Currently, the UK has the world's leading outsourcing of 51% of its offshore power plants. Denmark with 21% of the total offshore power plants in the world is in the second place. Other countries such as the USA, the Netherlands, Belgium, China, and Japan are also active in this field [4]. **Figure 1** shows the items that need to be considered in the market for the production and sale of electricity through ocean wind energy. The most critical issues regarding the wind ocean project are foundation type and water depth which are based on geographic information in the region.

1.1 Global ocean wind energy resource

1.1.1 Ocean wind energy in Asia

Over the past few decades, large countries like China and Japan have been using ocean wind energy for electricity production. According to wind energy reports at the World Wind Energy Council, China's investment in this area is more than the total European Union and is about 3.4 GW [5].

East Asia has a high potential for exploiting ocean wind energy, and many projects in this area, specifically in China, South Korea, and Japan, have been carried out and implemented. Among Asian countries, China has more shares in the use of wind energy. China was the first country in the Asia-Pacific region which used wind ocean energy. **Table 1** shows several offshore wind energy projects in Asia [5, 6].

China had installed more than 3.4 GW of ocean wind capacity at the end of 2016 and should end up with around 900 GW more by the end of 2030. According to

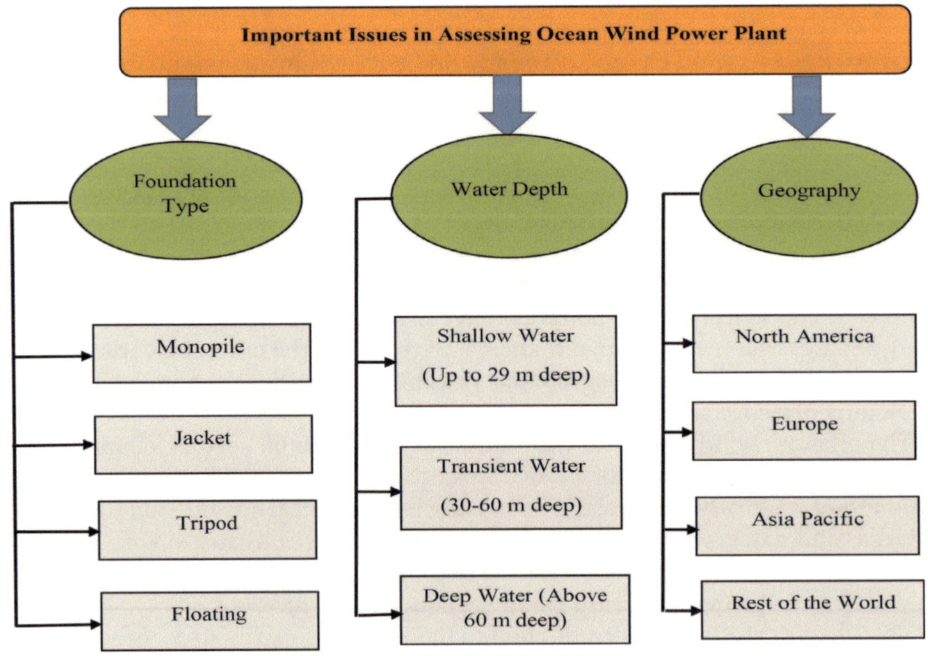

Figure 1.
Essential issues in assessing ocean wind power plant.

	Country	Planned capacity	Project name
1	China	1.5 MW	Bohai Suizhong, LiaoDong Bay
2	China	20 MW	Dongshan Island
3	China	50 MW	Hebei
4	China	100 MW	Nan'ao, southeast of Guangdong
5	China	25 MW	Shanghai Dong Mai
6	China	102 MW	Shanghai Dong Mai
7	China	100 MW	Fengxian No. 1
8	China	300 MW	Fengxian No. 2
9	China	400 MW	Nanhui
10	China	200 MW	Hengsha
11	Hong Kong	200 MW	Hong Kong offshore
12	Japan	2 × 600 KW	Setana, Hokkaido
13	South Korea	500 MW	Limjado, Jeonnam Province
14	Taiwan	4 MW	Ferry

Table 1.
Several ocean wind energy projects in Asia [5–7].

China's Five-Year Plan, five gigawatts will be added to the country's electricity grid by 2020 [4]. Taiwan also has the potential to produce electricity from ocean wind energy, and according to the plan of the Ministry of Energy, by 2025, it will be the top among the active countries in Asia that use this energy [7]. Today there are only two ocean turbines operating in the country, and two projects in a total of 320 MW are due to be installed by 2020. Japan has only 61 MW of installed energy for ocean wind energy by the end of 2016, given that it has more access to the ocean. South Korea has so far only mustered a couple of ocean wind prototypes and a single demonstration project, in a total of 35 MW [7].

1.1.2 Ocean wind energy in Europe

The European continent has many potentials for electric power generation through ocean wind energy. It is revealed that this energy will play an important role to produce electricity for Europe in the future [8].

The Netherlands started to work on offshore wind farms after Denmark [9]. Also, two ocean wind farms were built up in the Netherlands with two different capacity levels of 108 MW and 120 MW in 2006. Two new ocean wind farms were build up in Sweden by 2001 and 2002. Ireland constructed its first ocean wind farm in 2004 wind turbines of 3.5 MW. Moreover, Germany's first ocean wind farm was constructed with 20,000 MW capacity. The UK used ocean wind energy by 2000 with 3.8 MW capacity. France started to use ocean wind energy in 2005, but the construction of wind power plants for economic reasons was postponed to 2009. **Table 2** shows several offshore wind energy projects in Europe [8–10].

According to the EWEA[1], The European countries target is determining 20% of its power from sustainable sources by 2030. EWEA has set an objective to achieve 40 GW and 150 GW of ocean wind energy by 2020 and 2030, respectively. Additionally, through 2030, EWEA estimates yearly establishments of ocean

[1] European Wind Energy Association.

	Country	Planned capacity	Project name
1	Denmark	4.95 MW	Vindeby
2	The Netherlands	2 MW	Lely
3	Denmark	40 MW	Middelgrunden
4	Denmark	160 MW	Horns Rev
5	The UK	60 MW	North Hoyle
6	The UK	60 MW	Scroby Sands
7	The Netherlands	108 MW	Egmond ann Zee
8	Sweden	110 MW	Lillgrund
9	Netherland	120 MW	Princess Amalia
10	The UK	90 MW	Inner Dowsing
11	Germany	2.5 MW	Breitling
12	Ireland	25.2 MW	Arklow Bank
13	Sweden	10 MW	Yttre Stengrund
14	Italy	0.08 MW	Brindisi

Table 2.
Several offshore wind energy projects in Europe [8–10].

wind energy will be equal to 13,700 MW. Ocean wind energy will support 13.9% of total EU demand [10].

1.1.3 Ocean wind energy in Africa

Africa's wind energy resources are focused along the coastal area and mainland shelves. These regions ordinarily have high onshore and offshore wind energy possibilities. In 2004, the African Development Bank investigated to create a wind atlas of Africa and create a quantitative guide of wind speeds over the African continents [11]. Outcomes from the investigation showed that Africa's best wind energy is found in countries adjusted along the western, northern, eastern, and southern shores of the African continent. The special cases are landlocked countries such as Chad and Ethiopia where the topographical highlights of the land are responsible for the high wind speeds in some high-elevation zones. Additionally, according to research conducted in 2007, eight countries (Egypt, Somalia, Mauritania, Sudan, Libya, Chad, Kenya, and Madagascar) have high potential for onshore wind energy and five countries (Mozambique, Tanzania, Angola, South Africa, and Namibia) have high potential for ocean wind energy [11].

1.1.4 Ocean wind energy in America

The United States has vast ocean-wide areas such the Great Lakes, Hawaii, Alaska, and Gulf Coast with potential to use offshore wind energy to produce electricity [12]. As a result, the US Department of Energy's, Wind Energy Technologies Office has conducted many studies on various technologies to facilitate electricity generation from wind.

According to the US Department of Energy, the USA will have 3 GW, 22 GW, and 86 GW of ocean wind by 2020, 2030, and 2050, respectively. Therefore, the USA will utilize 5.5% of its accessible ocean wind resources. The US Bureau of Energy anticipated ocean wind improvement along both the Gulf of Mexico and

	Country	Planned capacity	Project name
1	The USA	30 MW	Block Island (RI)
2	The USA	468 MW	Cape Wind (MA)
3	The USA	500 MW	US Wind (MD)
4	The USA	1000 MW	DONG Energy (MA)
5	The USA	1000 MW	Deepwater ONE (RI/MA)
6	The USA	2000 MW	Dominion Virginia Power (VA)
7	The USA	450 MW	Blue Water's Mid-Atlantic Wind Park
8	The USA	400 MW	Offshore MW (MA)

Table 3.
Several ocean wind energy projects in the USA [12, 13].

West and East Coasts, in the Great Lakes, by 2050. **Table 3** shows several ocean wind energy projects in the USA [12, 13].

1.2 Ocean wind energy technologies

In the last few decades, the technology used to exploit ocean wind energy for generating electricity has been increased day by day. These technologies depend on the geographic region, the depth of water, and the wind speed. The main parts of the ocean wind power plant include [14]:

1. Tower

2. Blades

3. Gearbox

4. Power electronic components

5. Transmission system (cables)

6. Generator

1.2.1 Ocean wind turbine blade technology

The blades of the ocean wind turbine are one of the unique parts of the wind turbine structure. They have unique mechanical and aerodynamic characteristics. Moreover, the technology of manufacturing wind turbine blades has undergone new developments in both process fields and materials used in them. As a result, manufacturers of these blades are trying to optimize mechanical properties in the aerodynamic blades by design optimization and using new materials. Composite fibers and various resins, including various materials, have been used in the production of wind turbine blade rotor.

Extraction of kinetic energy from the wind is carried out by wind turbine blades. Therefore, having an optimal design to get the most energy out of the wind is very important. Wind turbine blade design consists of two main parts. In the first step, the aerodynamic design is performed to achieve the required power rating according to the turbine wind turbine and to obtain the highest electric power factor. In the second step, changes should be made to the blades so that the amount of aerodynamic noise generated by the blades is within the permissible range.

1.2.2 Ocean wind turbine tower technology

Wind turbine tower is the most significant, heaviest, and most expensive part of the wind turbine. Regarding the safety level, its failure can cause the entire wind turbine to fail. Proper design of the wind turbine tower significantly reduces the cost and increases the life of the wind turbine.

The tower is a cone-shaped steel structure with four segments mounted on each other by screws and flanges. The tower design includes the following main steps [15]:

- Connector analysis
- Shell strength analysis (static analysis, bending, and aging)
- Vibration analysis
- Design and selection of all internal components of the tower (entrance door, ladder, elevator, and internal platforms)

1.2.3 Ocean wind turbine gearbox technology

The purpose of using a gearbox is to transmit relatively large forces, change the torque or change the direction of rotation, or change the angle of the rotation axis. Gearboxes are increasing the nominal speed of a rotor from a small amount (a few tens of rpm) to a high value (at a rate of several hundred or several thousand rpm), which is suitable for triggering a standard generator. Ideally, the resultant value is constant in the torque at the inlet and outlet of the gearbox, but due to the energy losses in a mechanical device, torque is reduced in the output axis. In a wind turbine, the power transfer from the main rotor to the generator is usually done in three ways [16].

1.2.3.1 Direct drive transfer gearbox

In this method, the transmission is not used from the gearbox, and the torque is directly inputted from the main rotor to the generator. So, instead of using the gearbox and extending the main rotor, a generator with more poles is used. To accommodate more poles on the generator, the diameter should be increased. One of the benefits of using this design is to reduce the cost of the gearbox maintenance as well as reduce gearbox shocks and increase efficiency.

1.2.3.2 Power transfer by conventional gearbox (parallel shaft)

In this method, the power output is transmitted by a conventional gearbox to the generator. The gears used in this gearbox can be simple or spiral. To increase the upper period, it may be possible to use two or more rounds. In parallel shaft gearbox, the bearings are used to keep the gear shaft on the main body. In this type of gearbox, a helical gear is used, so in addition to radial force, the bearings must also bear a large axial force.

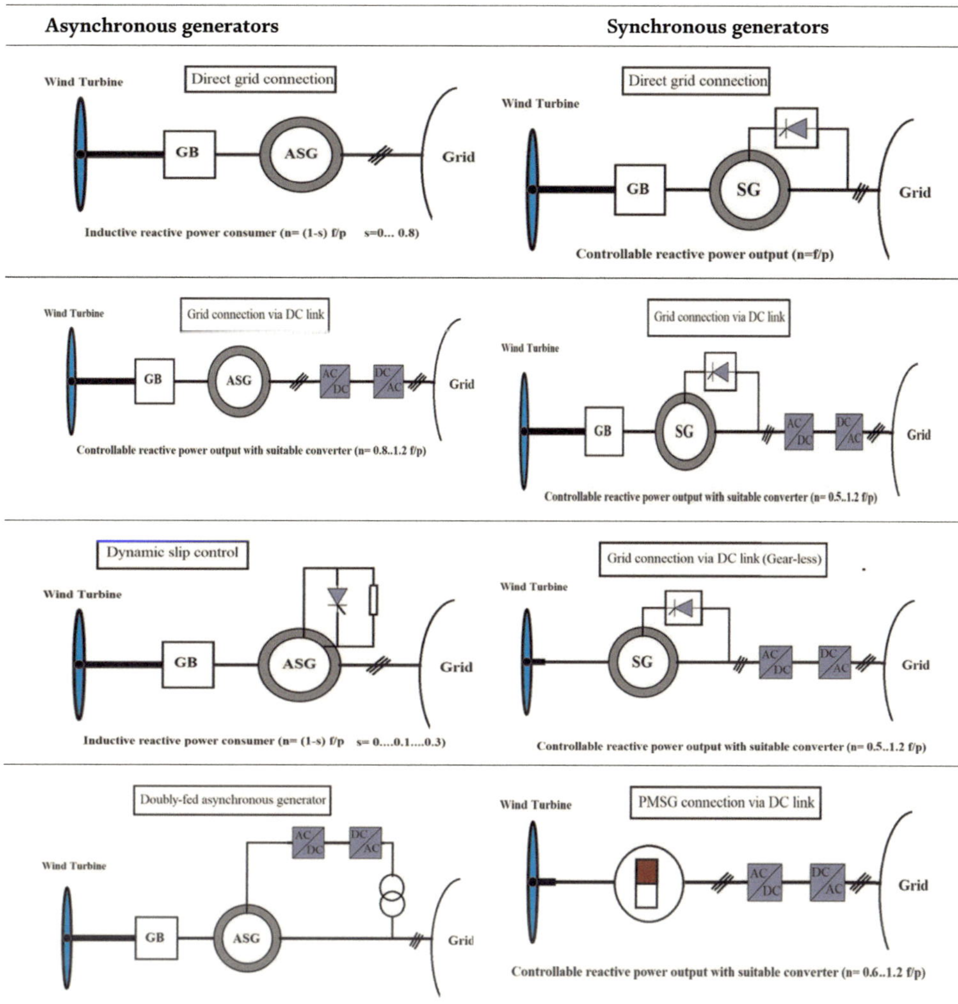

Table 4.
Different topologies of wind energy conversion systems.

1.2.3.3 Transmission by planetary gearbox

Using this type of gearbox is very common in wind turbines. The gearbox uses three types of gears, the Sun gear in the middle, the Planetary gear, and the Ring gear, which is an internal gear. The division of force into planetary planes, reduced gearbox size, reduced slip between the gear and the planet, and increased efficiency relative to other gearboxes are benefits of the planetary gearbox.

1.2.4 Ocean wind turbine energy conversion systems

Wind turbine blades convert wind energy into rotational energy in the transmission system, and in the next step, the generator transfers the turbine's energy to the grid. The most types of electric generator part in wind turbines are asynchronous and synchronous generators. Also, DC generators have been used for some smaller turbines. **Table 4** shows the different structure of ocean wind energy conversion [17].

In general, generators used to convert energy from offshore wind farms can be divided into two main categories, which can be described as follows:

- Synchronous generator
- Induction generator

    A. Squirrel cage induction generator

    B. Wound rotor induction generator

1.2.5 Ocean wind turbine power transmission technology

The construction of wind farms requires a large space. Therefore, the best option for removing this limitation is the construction of these power plants in the ocean. Because the distance between offshore wind farms and the distribution network is high, it is better to use high-voltage direct current (HVDC) to transmit energy produced. The suitable transmission for ocean wind farms based on HVDC is line-commutated HVDC and voltage source converter (VSC-HVDC) [18–20]. If the length of the transmission lines is less than 50 kilometers, the use of high voltage alternating current transmission systems is not recommended. In the HVDC

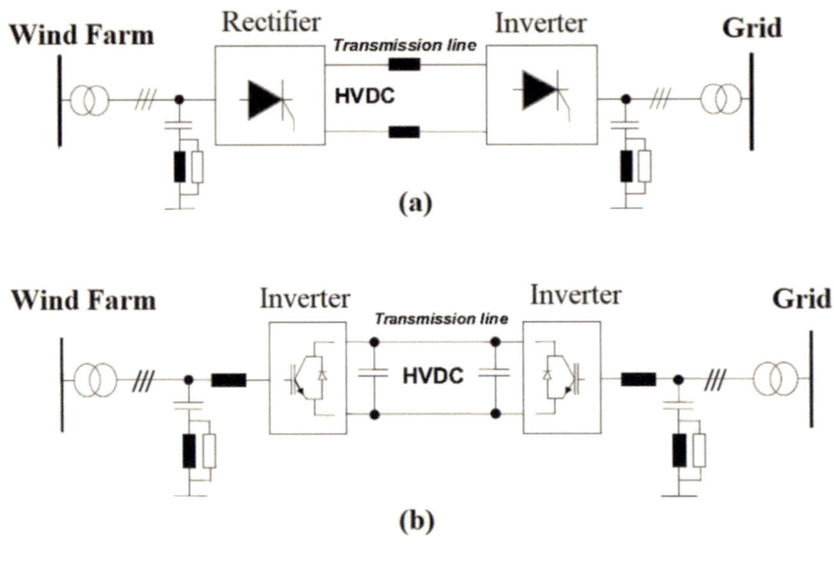

Figure 2.
(a) Structure of thyristor-HVDC and (b) structure of IGBT-HVDC.

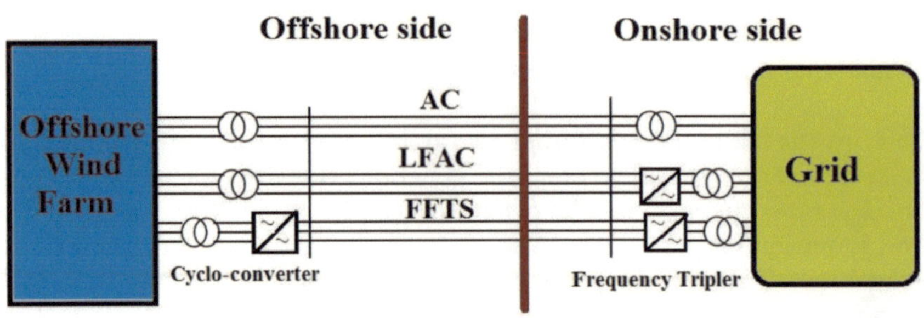

Figure 3.
Different types of AC transmission line of ocean wind farm.

technology to control active power, reactive power, and voltage thyristors is replacing with IGBT (**Figure 2a** and **b**).

The HVAC transmission network is divided into different types, each with its advantages and disadvantages. The low-frequency AC transmission (LFACT) and fractional frequency transmission system (FFTS) are new transmission systems which have been used for the wind farm as a solution to cover the disadvantages of conventional AC transmission line. **Figure 3** shows a different type of AC transmission line of the ocean wind farm.

2. Active and reactive power control in the ocean wind energy system

Recently, one of the most eco-friendly and accessible renewable energy sources (RESs) being utilized all across the world is wind energy [21]. Considering the beneficial characteristic of clean energy resources such as flexible control and regulation, RESs' expansion programs and sustainable growth to reduce the greenhouse gas emissions are the main purposes of European Commission in Energy Road map 2050 [22]. Regarding the necessity of employing renewable energies, and also the remarkable global growth in the use of such energy sources, wind energy has important advantages such as zero-emission energy production and low operating costs. In spite of these benefits, there is a severe uncertainty in predicting wind speed as a big challenge for such system's integration [23]. Various machines are used in wind turbines including permanent magnet synchronous generator (PMSG), squirrel cage induction generator (SCIG), and doubly fed induction generator (DFIG). Hence, the DFIG is one of the most significant types of generators being installed in wind turbines. In DFIG, both stator and rotor are connected to the main power grid directly and by power electronic converters, respectively [24–28]. As it can be seen in **Figure 1**, the typical circuit of a doubly fed induction generator DFIG is specified, by taking into consideration the several important parts including maximum power point tracking (MPPT), rotor and grid side controllers and power electronic converters, and pulse wide modulation PWM. Both windings of the stator and rotor of the induction machine are connected to the grid directly and by power converters, respectively. To active and reactive power control, several different types of controller have been evaluated in pervious, research works. For instance, Ref. [29] uses sliding mode (SM) and PI controller, for controlling the stability and also to track reference power and remove fluctuations or active and reactive powers' disturbances. After pointing out the performance of the PI controller in the output, the obtained results are compared with a SM controller. It has been revealed that the PI controller has a more desired performance than SM controller from the output responses of the controllers' point of view. In [30], a neural-type-1 fuzzy controller is used to produced powers of wind turbine; the derived results are compared to a PI controller, specifying that the neural-fuzzy controller operates better than the PI controller. Refs. [31, 32] have used fuzzy-PI and sliding mode and also robust fuzzy-sliding mode controller (F-SMC) to control and better manage the generated Ps and Qs of the wind turbine system output. Hence, comparing of controllers' performance indicates that the output responses of the fuzzy-proportional integration controller are better than sliding mode controller. All the controllers used in the literature have been used to enhance the stability and to eliminate the fluctuation and disturbances as well as improve the reference power tracking. Furthermore, each control method has modified the pervious approaches and also improved the active and reactive powers extraction of DFIG. Another type of controller employed in DFIG generator is the fuzzy-sliding mode controller. In fact, the F-SMC is a controller which finds the best numerical values for the scaling factor of the sliding mode. Due to the presence of fuzzy controller, all gain coefficient value is more accurate. Wind turbine operation is affected by uncertainty due to the uncertainty in the wind speed forecasts; thus the abovementioned controllers are not appropriate as they are not capable of dealing with severe uncertainties. It is noteworthy that the PI controller is point to point and type-1 fuzzy controller is able to only cover a small range of uncertainty in its outputs. By considering the high uncertainty in the wind speed, it is needed to design a controller with the capability of covering severe uncertainties. According to the control methods used in the literature, and also in order to improve and enhance the power control (Ps & Qs) in wind turbines systems, a new control method based on type-2 fuzzy logic control laws is presented. In this research, all parameters and equations are linear enabling the system designer to use MAMDANI inference system. The T2FL controller is a suitable alternative for the controlling of the powers in DFIG to deal with severe uncertainty.

2.1 Mathematical model of wind energy conversion system

Indeed, the DFIG is an induction machine which its stator and rotor is interfaced to the main power grid. The connection of the stator and electrical energy transmission from its windings to the main power grid is carried out by 3-phase power transmission lines from wind turbine to the grid. On the other hand, the rotor wounds of the induction machine are fed through the AC-DC/DC-AC back to back power converters, which received the electrical energy from 3-phase power transmission lines between stator and the main power grid. In this section, a general model is presented within a dynamic framework, by considering the variation parameters of the wind turbine based on induction machine. In this regard, the mathematical and dynamic equations are expressed as currents, voltages, and flux relations in q-d-0 reference frame and also the electromagnetic and mechanical torque. **Figure 1** illustrates the wind turbine circuit loop general operation process with controllers, in particular the rotor-side one [30, 33].

2.1.1 Universal model of the wind turbine

By taking into consideration the rate of wind speed at the different times and also due to the coefficient $(\lambda_{wind}.\beta)$, the mechanical power transferred from wind turbine blades to the rotor shaft can be written as follows [27, 34]:

$$P_t = \frac{\pi\, \rho\, R^2 V_{wind}^3\, Cp\,(\lambda_{wind}.\beta)}{2} \tag{1}$$

wherein ρ is air density and R, V_{wind}, and$Cp\ (\lambda_{wind}.\beta)$ are the radius of the turbine rotor, wind speed, and the coefficient of the power capture, respectively. Hence, the power capture coefficient $Cp(\lambda_{wind}.\beta)$ can be defined as follows [33, 34]:

$$Cp\,(\lambda_{wind}.\beta) = C_1\left[\frac{C_2}{\lambda_i}C_3\beta - C_4\right]e^{\frac{-C_5}{\lambda_i}} + -C_6\lambda \tag{2}$$

where λ_{wind} and β are denoted as ratio of the tip speed and blades pitch angle, respectively. Therefore, λ_{wind} can be calculated according to the following:

$$\lambda_{wind} = \frac{\omega_{total}.R}{U} = \frac{\omega_r.G.R}{U} \tag{3}$$

With regard to the relation (1), the input mechanical torque of the wind turbine is obtained by following:

$$T_{mech} = \frac{P_{total}}{\omega_{total}} = \frac{\frac{\pi\,\rho\,R^2 V^3_{wind}\,Cp\,(\lambda_{wind}.\beta)}{2}}{\omega_{total}} \tag{4}$$

2.1.2 DFIG mathematical model theory

Generally, there are three states or significant vector models, which are used to design an induction machine called as **Park**, **Clark**, and **Concordia**. Hence, 2-phase vector model reference frame of PARK is considered for designing the induction generator. On the other hand, all relationships of doubly fed induction machine are used, is under the dynamic equations in 3-phase (a, b, c) transformation frame system to (q-d-0) 2-phase reference frame. The main parameters of formulas can be described as stator and rotor voltage, current, flux in q-d-0 system, and also electromagnetic and mechanical torque [35–38].

Figures 4 and **5** represent the vector diagram of the DFIG PARK's model and 2-phase reference frame, respectively. In these figures the conversion system from (a, b, c) reference rotating frame to (q-d-0) reference frame has been shown [32].

DFIG's voltage equations of the stator and rotor in 2-phase reference system are defined as follows [34, 37]:

$$\begin{cases} U_{d\ Stator} = R_{Stator} I_{d\ Stator} + \dfrac{1}{\omega_{base}} \dfrac{d\psi_{d\ Stator}}{dt} - \dfrac{\omega}{\omega_{base}} \psi_{q\ Stator} \\ U_{q\ Stator} = R_{Stator} I_{q\ Stator} + \dfrac{1}{\omega_{base}} \dfrac{d\psi_{q\ Stator}}{dt} + \dfrac{\omega}{\omega_{base}} \psi_{d\ Stator} \end{cases} \tag{5}$$

$$\begin{cases} U_{d\ Rotor} = R_{Rotor} I_{d\ Rotor} + \dfrac{1}{\omega_{base}} \dfrac{d\psi_{d\ Rotor}}{dt} - \dfrac{\omega - \omega_{Rotor}}{\omega_{base}} \psi_{q\ Rotor} \\ U_{d\ Rotor} = R_{Rotor} I_{q\ Rotor} + \dfrac{1}{\omega_{base}} \dfrac{d\psi_{q\ Rotor}}{dt} + \dfrac{\omega - \omega_{Rotor}}{\omega_{base}} \psi_{d\ Rotor} \end{cases} \tag{6}$$

where U.R.I.ψ is characterized as the stator and rotor voltages, resistances, and currents, fluxes in d-q reference frame, and also the rotor base speed, rotor speed, and angular speed is denoted by $\omega_b.\omega_r.\omega$ respectively.

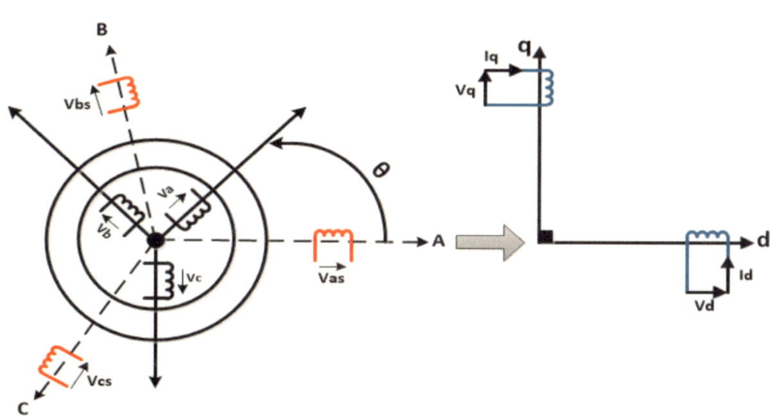

Figure 4.
Doubly fed induction generator PARK's model [32].

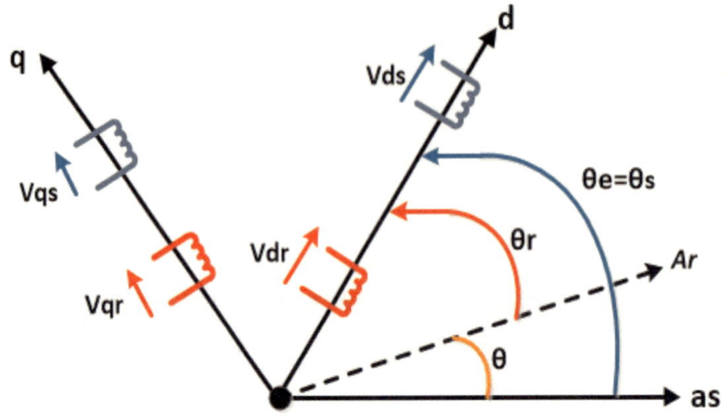

Figure 5.
The DFIG in the 2-phase reference frame [32].

Also, the induction machine's currents are in 2-phase reference frame, which are calculated as follows:

$$\begin{cases} \mathrm{I_{d\ Stator}} = \dfrac{\Psi_{\mathrm{d\ Stator}}}{\mathrm{X_{l\ Stator}}} - \dfrac{\Psi_{\mathrm{md}}}{\mathrm{X_{l\ Stator}}} \\ \mathrm{I_{q\ Stator}} = \dfrac{\Psi_{\mathrm{q\ Stator}}}{\mathrm{X_{l\ Stator}}} - \dfrac{\Psi_{\mathrm{mq}}}{\mathrm{X_{l\ Stator}}} \end{cases} \tag{7}$$

$$\begin{cases} \mathrm{I_{d\ Rotor}} = \dfrac{\Psi_{\mathrm{d\ Rotor}}}{\mathrm{X_{l\ Rotor}}} - \dfrac{\Psi_{\mathrm{md}}}{\mathrm{X_{l\ Rotor}}} \\ \mathrm{I_{q\ Rotor}} = \dfrac{\Psi_{\mathrm{q\ Rotor}}}{\mathrm{X_{l\ Rotor}}} - \dfrac{\Psi_{\mathrm{mq}}}{\mathrm{X_{l\ Rotor}}} \end{cases} \tag{8}$$

By considering current Eqs. (7) and (8), the parameters $\psi.\psi_{Magnetic}.X_{l\ Stator}.X_{l\ Rotor}$ are known as the, flux, magnetizing flux, and stator and rotor leakage reactance in q-d two-dimensional vector space. Hence, under such condition the fluxes of the stator and rotor windings can be formulated as follows:

$$\begin{cases} \Psi_{\mathrm{q\ Stator}} = \omega_{\mathrm{base}} \displaystyle\int \left(\dfrac{\mathrm{R_{Stator}}\left(\Psi_{\mathrm{mq}} - \Psi_{\mathrm{q\ Stator}}\right)}{\mathrm{X_{l\ stator}}} + \mathrm{V_{q\ Stator}} \right) \mathrm{dt} \\ \Psi_{\mathrm{d\ Stator}} = \omega_{\mathrm{base}} \displaystyle\int \left(\dfrac{\mathrm{R_{Stator}}\left(\Psi_{\mathrm{md}} - \Psi_{\mathrm{d\ Stator}}\right)}{\mathrm{X_{l\ Stator}}} + \mathrm{V_{d\ Stator}} \right) \mathrm{dt} \end{cases} \tag{9}$$

$$\begin{cases} \Psi_{\mathrm{q\ Rotor}} = \omega_{\mathrm{base}} \displaystyle\int \left(\dfrac{\mathrm{R_{Rotor}}\left(\Psi_{\mathrm{mq}} - \Psi_{\mathrm{q\ Rotor}}\right)}{\mathrm{X_{l\ rotor}}} + \dfrac{\omega_{\mathrm{Rotor}}}{\omega_{\mathrm{base}}} \Psi_{\mathrm{d\ Rotor}} + \mathrm{V_{q\ Rotor}} \right) \mathrm{dt} \\ \Psi_{\mathrm{d\ Rotor}} = \omega_{\mathrm{base}} \displaystyle\int \left(\dfrac{\mathrm{R_{Rotor}}\left(\Psi_{\mathrm{md}} - \Psi_{\mathrm{d\ Rotor}}\right)}{\mathrm{X_{l\ rotor}}} + \dfrac{\omega_{\mathrm{Rotor}}}{\omega_{\mathrm{base}}} \Psi_{\mathrm{q\ Rotor}} - \mathrm{V_{d\ Rotor}} \right) \mathrm{dt} \end{cases} \tag{10}$$

In addition, by taking into consideration the current and flux parameters in 2-phase reference frames of the rotor and stator of the induction generator, the torques can be obtained as (11) and (12):

$$T_{electromagnetic} = \frac{1.5\,P\left(\psi_{d\ Stator} I_{q\ Stator} - \psi_{q\ Stator} I_{d\ Stator}\right)}{2\omega_{base}} \tag{11}$$

where P, ω_{base} are the number of pole and base speed of the induction generator's rotor, respectively, which are considered in the calculation of electromagnetic torque. Due to the relation (11), the mechanical torque can be written as follows:

$$T_{Mechanical} = J\frac{d}{dt}\omega_{rm} - T_{em} + T_{damp} \tag{12}$$

In (12), the parameters T_{damp}, J are defined as the damping torque and rotor moment of inertia, considering the rotor speed variations.

2.1.3 P_s & Q_s power control process in DFIG-based WECS

The active and reactive power control strategy is that the fuzzy controller by receiving error signals and derivative of the error and also, after a control process, the signals generated through the controller is delivered to the power electronic converters and DFIG's rotor and to the active and reactive output power control of the wind turbine. The relationships of P_s & Q_s powers are defined as follows [40, 41]:

$$P_s = \left[\frac{1}{X_{l\ Magnetic}}(\psi_{d\ Stator}) + \frac{X_{Magnetic}}{X_{l\ Stator}}(1.5\, I_{d\ Rotor}\, V_{d\ Rotor})\right] \tag{13}$$

$$Q_s = \left[\frac{1}{X_{l\ Magnetic}}\left(\psi_{q\ Stator}\right) + \frac{X_{Magnetic}}{X_{l\ Stator}}\left(1.5\, I_{q\ Rotor}\, V_{q\ Stator}\right)\right] \tag{14}$$

2.2 Controller design based on the type-2 fuzzy logic theory T-2FLC

2.2.1 Type-2 fuzzy controller statement

T-2 fuzzy controller is a developed controller in that its operation strategy is under the uncertainty. Therefore, this type of controller is appropriate for systems with high uncertainty such as wind or solar power plants, which the generation is exactly under the pure uncertainty. The performance of the T2FLC because of covering a large scale of high uncertainty to control the wind turbine parameters is more desired than the T1 fuzzy or another controller technique. Structurally, both T1 and T2 fuzzy controllers are the same, but with this difference in the interior structure of T2FLC, due to the presence of uncertainty, there is a section as type reducer (TR). The calculation and the conversion of the type of fuzzy from type 2 to type 1 are the important functions of the TR section. Depending upon the linearity and nonlinearity of the model, two types of inference system exist in T2FLC. The MAMDANI inference system is used for the systems with linear equations, and TSK inference system is employed for the systems with nonlinear equations.

2.2.2 Extended fuzzy sets

Generally, T2 fuzzy sets are extended of type-1 fuzzy sets. On the other hand, T2FS is a fuzzy set with membership degree of fuzzy. Type-2 fuzzy sets can compensate the limitation of type-1 fuzzy sets in covering uncertainties as a new method with its specific advantages. Forasmuch as fuzzy sets are defined based on linguistic variables, thus the T2FS is appropriate to model uncertainty process using linguistic variable. The primary membership grade in T1FSs is a crisp number in [0, 1], whereas the primary membership grade in T2FSs is a T1FSs in range of

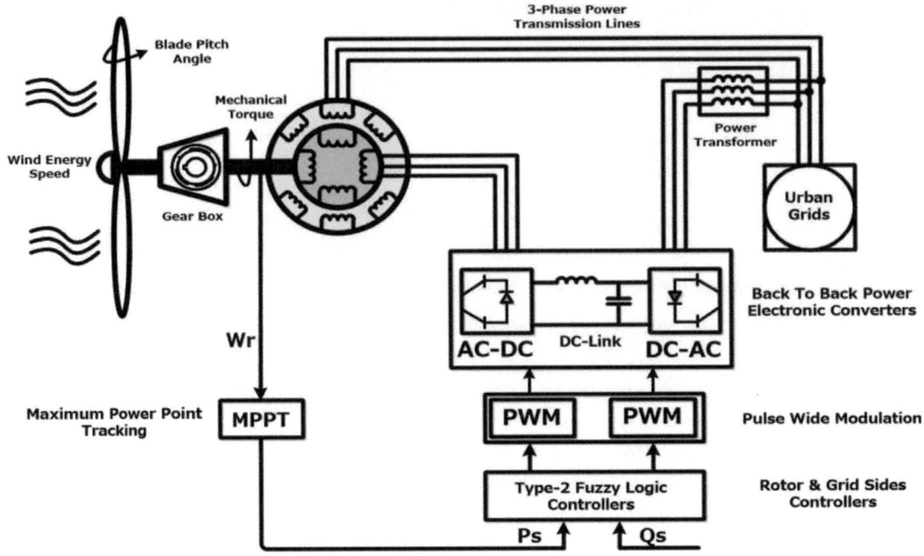

Figure 6.
The general circuit of a DFIG and its connection to the grid.

[0, 1] and the secondary membership is a crisp number in [0, 1] as well [17, 22]. **Figure 6** represents T2FSs with its upper, lower, and footprint ranges of the uncertainty.

As shown in **Figure** 7, A T2FS consists of foot printing of uncertainty (FOU), upper membership function (UMF), lower membership function (LMF), and embedded fuzzy system wherein the FOU and embedded FS have been shown as blue lines. If all membership grades of FT2 sets are the same in the secondary part of FT2 sets, the sets of FT2 are the internal type, otherwise the general type. With respect to the uncertainty in the membership function of FT2 sets, the general concept of type-2 fuzzy set is defined by the relations below [37, 38].

A type-2 fuzzy set is characterized in a function (H) and is described as follows:

$$H = \{[(y.v).\mu_H(y.v)]\}.\forall_y \in Y.\forall_v \in Z_y \subseteq [0.1] \tag{15}$$

wherein

$$y \in Y.v \in Z_y \subseteq (0.1).0 \leq \mu_H\ (y.v) \leq 1\,.$$

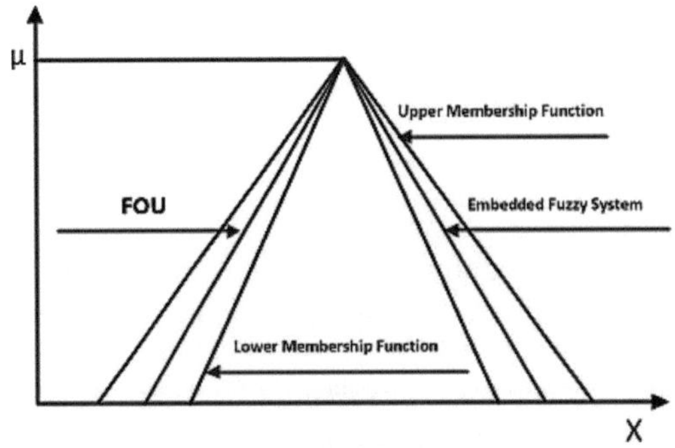

Figure 7.
Type-2 fuzzy set with FOU and embedded FS and lower and upper memberships.

In order to reduce the computational complexity, the interval fuzzy sets (IFSs) are proposed as an alternative to the general fuzzy sets (GFSs). Hence, a set $(\widetilde{H})$ of interval type2 is defined as follows:

$$\begin{cases} \widetilde{H} = \int_{y\in Y}\int_{v\in Z_y} \frac{1}{(y.v)} Z_y \subseteq [0.1] \\ \widetilde{H} : Y \to \{[E.F] : 0 \le E \le F \le 1\} \end{cases} . \tag{16}$$

The performance of fuzzy inference engine is under the considered rules. With seven membership functions, 49 of the commands written to form IF-THEN have been considered as represented in **Table 1** [37–42]. The relation of footprint of uncertainty is as follows:

$$\text{FootPrintin}\left(\widetilde{H}\right) = \cup_{\forall y \in Y} Z_y = \{[\,y.v\,] : \cup \in Z_y \subseteq [0.1]\} \tag{17}$$

According to the impacts of the uncertainty in FT2 sets, the bound of FT2 sets includes two fuzzy type-1 set membership functions as upper membership function (UMF) and lower membership function LMF. The embedded fuzzy sets in the set of $\boldsymbol{H}$ can be defined as follows as a $\widetilde{\boldsymbol{H}}_e$ set [18]:

$$\widetilde{H}_e = \int_{y\in Y} \frac{\left[\frac{1}{v}\right]}{y} .v \in Z_y \tag{18}$$

2.2.3 Principle process of the type-2 fuzzy logic inference system

The general performance of T2FLC is based on rules and relationships which have been considered for it. As it can be seen in **Figure 8**, the principle process of type-2 and type-1 fuzzy control systems are the same, but with this difference that the FT2 control system has a unit called fuzzy type reducer. FT2 system comprises five important parts in that the first part is the fuzzifier unit, while the inputs of [0, 1] interval are converted to fuzzy sets. The second part is the inference engine unit wherein; all fuzzy sets are inferred by rule base unit, simultaneously. Depending upon linearity or nonlinearity of the fuzzy control inputs and the equations, the inference system in (T2FLS) control system can be MAMDANI or TSK. The next part after the inference engine taken into consideration the most important part of FT2 logic system is the rule base unit. All fuzzy inference calculations are according to the human knowledge and written in the frame of IF-THEN.

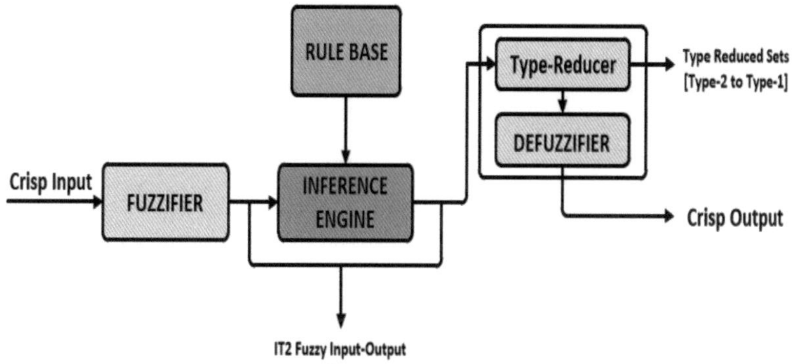

Figure 8.
Overall process of FT2 logic system.

The fourth section of FT2 system is the type reducer. Since the FT2 sets are based on the uncertainty and due to the high computational burden of the fuzzy system, this is not possible for the system output to be converted to [0, 1] directly. First, all the sets of the FT2 are converted to FT1 sets using the type reducer and then applied to the defuzzifier unit and converted to [0, 1] in the output. The last part of the FT2 system is the defuzzifier unit in that its performance is in the opposite way of the fuzzifier system and converts all fuzzy sets to [0, 1] [38–40].

By taking into consideration the significant parts of type-2 fuzzy topology, the process of inference is expressed in the form of mathematical, which is denoted as follows:

$$\mathrm{E^M} = \left[\mathrm{F_{Lower}.F_{Upper}}\right] \tag{19}$$

subject to

$$\mathrm{F^M_{Lower}} = \min\left[\left(\mu_{\mathrm{N}^K_1}\right)_{\mathrm{Lower}}(\mathrm{y}_1)\dots\left(\mu_{\mathrm{N}^K_{\mathcal{P}}}\right)_{\mathrm{Lower}}(\mathrm{y}_{\mathcal{P}})\right] \tag{20}$$

$$\mathrm{F^M_{Upper}} = \min\left[\left(\mu_{\mathrm{N}^K_1}\right)_{\mathrm{Upper}}(\mathrm{y}_1)\dots\left(\mu_{\mathrm{N}^K_{\mathcal{P}}}\right)_{\mathrm{Upper}}(\mathrm{y}_{\mathcal{P}})\right] \tag{21}$$

And the minimum and maximum computational of type reducer can be expressed in the fractional functions as follows:

$$Z_l = \min_{\delta_i \in \left[F^K_{Lower}(x_i)\cdot F^K_{Upper}(x_i)\right]} \frac{\sum_{i=1}^{N} x_i\,\delta_i}{\sum_{i=1}^{N} \delta_i} \tag{22}$$

$$Z_r = \max_{\delta_i \in \left[F^K_{Lower}(x_i)\cdot F^K_{Upper}(x_i)\right]} \frac{\sum_{i=1}^{N} x_i\,\delta_i}{\sum_{i=1}^{N} \delta_i} \tag{23}$$

Finally, defuzzification is the next step after the type reduction unit which in order to achieve the controller's output is done by:

$$\mathrm{Z_c} = \frac{\mathrm{Z_l}}{2} + \frac{\mathrm{Z_r}}{2} \tag{24}$$

2.2.4 Type-2 fuzzy controller design

Generally, **Figure 9** shows the process of the data inference, analyze and conversion them from crisp system input [0, 1] to the type-2 fuzzy system and again transform to the crisp system output [0, 1]. According to the main system's input equations, the design of FT2 controller has been done using the FT2 toolbox. The FT2 controller detail, such as error, change of error that is gain input (KP, KD), fuzzy inference system unit, output gain (KU) with its intervals, the number of considered membership functions for inputs and output, some of its laws, and, also, the type of inference, is expressed in the form of a toolbox. Given the linearity of the equations of DFIG, the MAMDANI inference system with Gaussian membership functions has been considered for the FT2 controller [39]. The main part of type-2 controller is the fuzzy inference (FIS) section, in which all operating levels of fuzzy sets can be done by this part. Since this work is focused on the rotor-side controller RSC and also by considering the presence of uncertainty in the wind speed, elimination of the oscillation (overshot), as well as stability enhancement of the output powers Ps & Qs of the wind turbine based on DFIG, is the principal target of this essay. With regard to **Figure 1** that indicated the general structure of

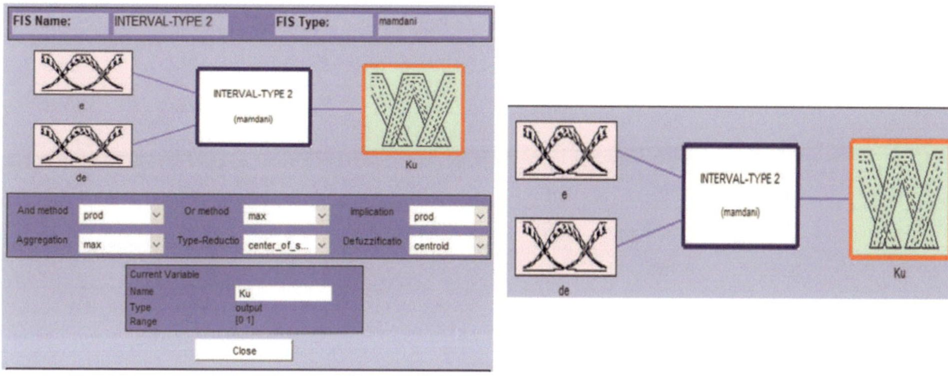

Figure 9.
Type-2 fuzzy logic toolbox.

the DFIG, hence, as it can be seen, the type-2 fuzzy controller input parameters are the active and reactive powers, and its output is the voltage of the rotor in the q-d-0 reference frame. The performance of fuzzy inference engine is under the considered rules. With seven membership functions, 49 of the commands written to form IF-THEN have been considered as represented in **Table 1** [37, 38, 40]. To better indicate the concept of the type-2 toolbox and its application to T2FL controller design, the general scheme of an interval type-2 toolbox with MAMDANI inference strategy and also the type of inputs and output membership functions with its ranges are depicted in **Figures 5** and 7, respectively. As shown in **Figure 8**, the structure of type-2 fuzzy logic controller is composed of input gains (KP, KD), type-2 fuzzy inference unit, output gain (KU), and plant as well. Indeed, the plant section is a mathematical transfer function. Since the type-2 fuzzy inference section is the main part of T2 controller and on the other hand its function is directly based on the IT2 toolbox, thus, under such conditions, it is required that all parameters' information about the type-2 controller system, such as input and output scaling factors, number of rules and command, membership functions, and its ranges, are defined in the toolbox as well. All sections of the T2 fuzzy logic controller are shown in **Figure 10**.

Notation 1: Each letter in **Table 5** has a special meaning. For instance, negative big (NB), negative medium (NM), negative small (NS), zero (ZO), positive small (PS), positive medium (PM), and positive big (PB).

As shown in **Figure 11**, the structure of type-2 fuzzy logic controller is composed of input gains (KP, KD), type-2 fuzzy inference unit, output gain (KU), and plant as

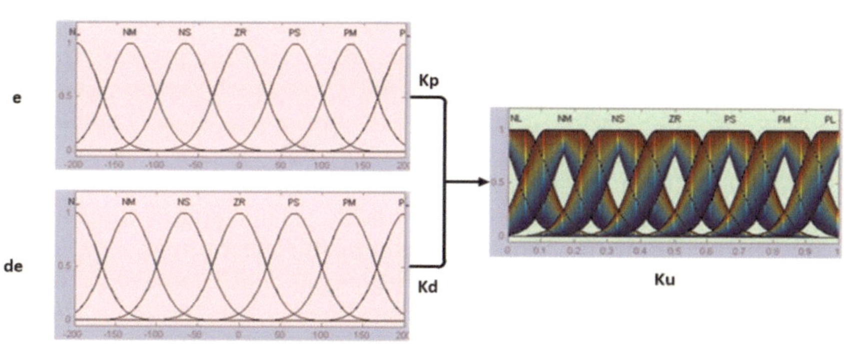

Figure 10.
Input and output Gaussian membership functions with its intervals.

E/COE	**NB**	**NM**	**NS**	**ZO**	**PS**	**PM**	**PB**
NB	NB	NB	NB	NM	ZERO	ZERO	ZERO
NM	NB	NB	NB	NM	ZERO	ZERO	ZERO
NS	NB	NB	NB	NM	ZERO	ZERO	ZERO
ZO	NM	NM	NM	ZERO	PM	PM	PM
PS	ZERO	ZERO	ZERO	PB	PB	PB	PB
PM	NB	NM	NS	PM	PB	PB	PB
PB	PS	PM	PB	PB	PM	PB	PB

Table 5.
Type-2 fuzzy rule chart.

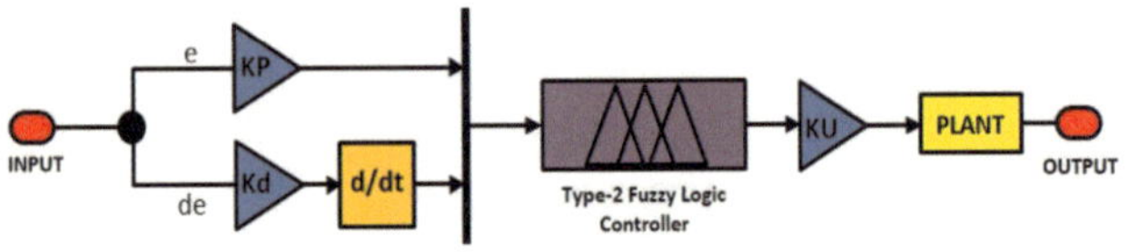

Figure 11.
The main structure of the type-2 fuzzy logic controller simulation.

well. Indeed, the plant section is a mathematical transfer function. Since the type-2 fuzzy inference section is the main part of T2 controller and on the other hand its function is directly based on the IT2 toolbox, thus, under such conditions, it is required that all parameters' information about the type-2 controller system, such as input and output scaling factors, number of rules and command, membership functions, and its ranges, are defined in the toolbox as well. All sections of the T2 fuzzy logic controller are shown in **Figure 11**.

Each of letters in **Table 1** has a special meaning. For example, negative big is the meaning of NB, and ZO is the abbreviation of zero, while the following describes the fuzzy rules:

If error is **negative big** and change of error is **negative big,** then KU is **negative big**.

2.2.5 Tuning of FT2 controller's gains using the PSO algorithm

PSO is one of the most popular optimization algorithms which is operated according to the social treatment of birds and aquatics movement. The process of optimization in the algorithm ends whenever using the pre-defined stop criteria [43, 44]. In this article, (PSO) algorithm is used to tuning the input and output scaling factors of the controller. To optimize the output powers (Ps & Qs) of the wind turbine through the T2 fuzzy controller, it is required to properly tune the input and output gains of the controller [45–47]. Under such conditions, each of the input and output scaling factors of the type-2 controller will have a suitable number, in which its numerical amounts are determined by PSO algorithm. In the presence of uncertainty and due to the complexity and the large number of the FT2 equations, it would be very difficult or even impossible to choose an optimal number or enter values manually into the input and output gains. Accordingly, PSO algorithm has been used in this paper to accelerate adjusting the coefficients to get the proper number and more accurate response to regulate the input and output scaling factors of the controller. The PSO algorithm is based on the particles' behavior including the velocity and the location of particles [48–53]. Taking into consideration the general structure of the PSO algorithm, the process of coefficients regulation of the

FT2 controller's input and output gains using PSO algorithm is defined in three steps. At the first step, a general cost function is created including the names of the controller's gains characteristic; the name of the main system that the type-2 fuzzy controller is considered for, i.e., a DFIG-based wind turbine; and the sum of error and the change of the error. In the second step, the main values such as the number of parameters, the minimum and maximum values of the input and output of the FT2 controller gains (KP, KD, KU), the name of the cost function, the number of maximum iteration, as well as all parameters relating to the PSO algorithm are defined. In the third step, the best numerical value of the FT2 controller gains is determined by running the PSO algorithm. By considering a larger number of iteration loops in the algorithm to adjust the gains of the controller, the output response will be improved. The general structure process of the type-2 controller's gain regulation has been depicted in **Figure 12**. As shown in **Figure 9** and also with the presence of the T2 fuzzy controller in this system, the PSO algorithm adjusts all scaling factors of the T2FL controller by receiving the error and change of error (E, COE) as the input and then chooses the best value for each gain of the controller (KP, KD, KU) in the output. To better understand the optimization procedure by the PSO algorithm, all the algorithms' steps are described as a flowchart in **Figure 13**.

Notation 2: Indeed, the PSO algorithm is based on the cost function for which it is intended. In order to membership functions tuning of the type-2 fuzzy controller gains, the cost function is defined as follows:

```
Function H= Cost Function-FCN (KP, KD, KU)
Sim ('DFIG')
H=Sum ((e. ^ 2) + (De. ^ 2))
End
```

2.3 Simulation results

This part expresses the simulation results obtained using the presented framework. In this regard, the obtained results using the proposed FT2 controller are compared with those obtained by the FT1 controller. **Figure 14(a)** and **(b)** shows the error and change of error surfaces of the FT1 and FT2 controllers, respectively. In this regard, the FT2 controller has a smoother surface than FT1 due to the covering uncertainty in a large and different ranges and high computational burden. This paper is focused on the power (P & Q) control using the RSC. According to the circuit loop of doubly fed induction machine, the general power control process in

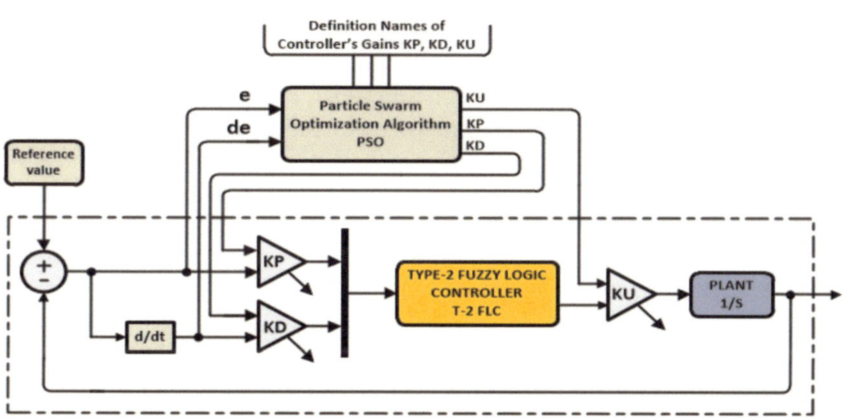

Figure 12.
Tuning of fuzzy type-2 controller gain process using PSO algorithm.

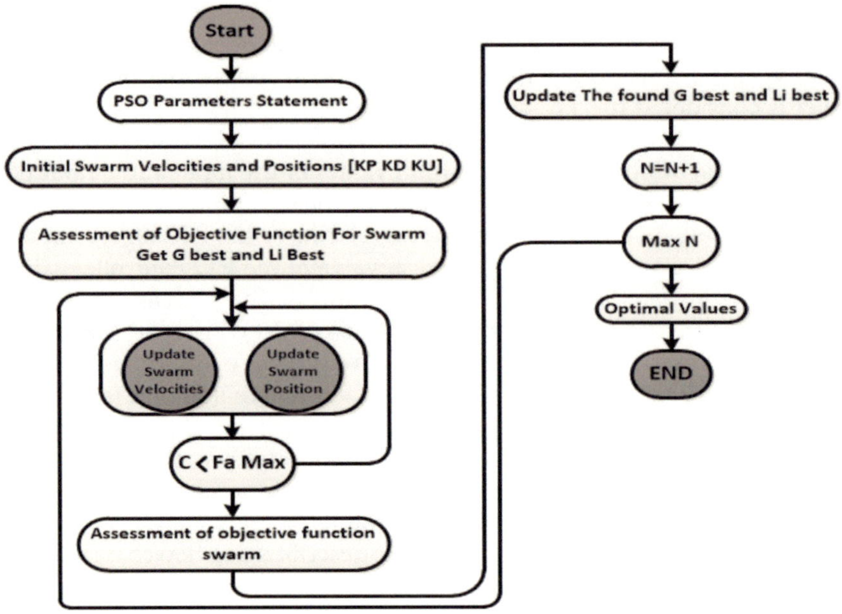

Figure 13.
Flowchart of the PSO algorithm.

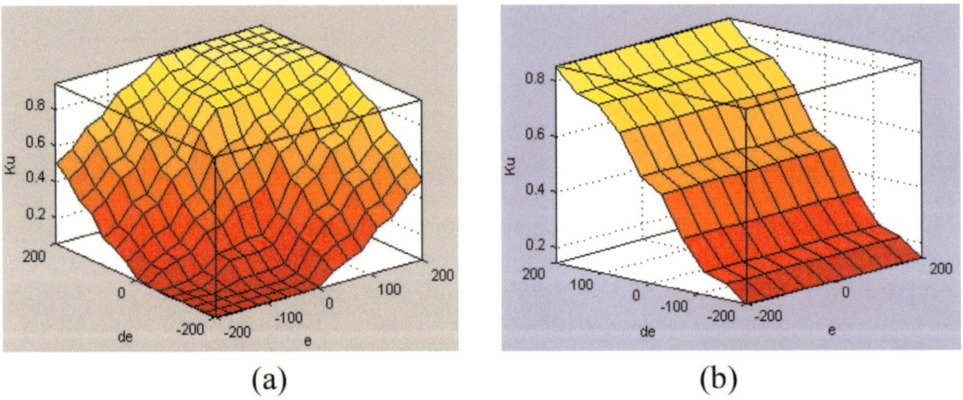

Figure 14.
(a) and (b), The FT1 and FT2 control surface.

the DFIG-based wind turbine can be stated in multiple stages. At the first stage, after entering the value of the measured power (the generated power by the DFIG in the initial moment without controller's function), reference power is compared by the type-2 fuzzy controller. Regarding the transfer function considered for the T2 FLC, the output signal of the controller is the rotor voltage in *d-q* reference frame. Since the input of pulse wide modulation (PWM) unit is the voltage in the *a*, *b*, *c* reference frame, at the second stage, first the controller's output voltage in *d-q* frame is converted to the a, b, c frame by a *d-q* to *a*, *b*, *c* unit transformation; after that the controller's output signal will be sent to the PWM block, and at the third stage, the output signal of the PWM will be transferred to the rotor-side power converter. In the presence of uncertainty of the wind speed, the main goal of this simulation is to show the stability of the powers on the considered reference numerical amount, using a T2 fuzzy controller. In order to power stability on the value of 400 W, the reference power should be adjusted to 400. The active and reactive power output responses have been exactly stabilized at the reference amount which is considered for the outputs of the T1 and T2 fuzzy controllers. All the results are depicted in **Figures 15** and **16**. As it can be seen in the figures, both the powers Ps & Qs are in the stable mode after multiple overshot in the transient state of the FT1 controller output, but in FT2 controller, the active and reactive powers have been stabilized without any overshoot or oscillation in the transient state. Indeed, before the active power and reactive power are stabilized in the output, all overshoots or disturbances are removed in the transient states by the FT2 controller. In the FT2 controller due to the high computational burden, the active and reactive powers become stabilized with more latency compared to the FT1 controller. In this part, to better indicate the wind turbine and also for more

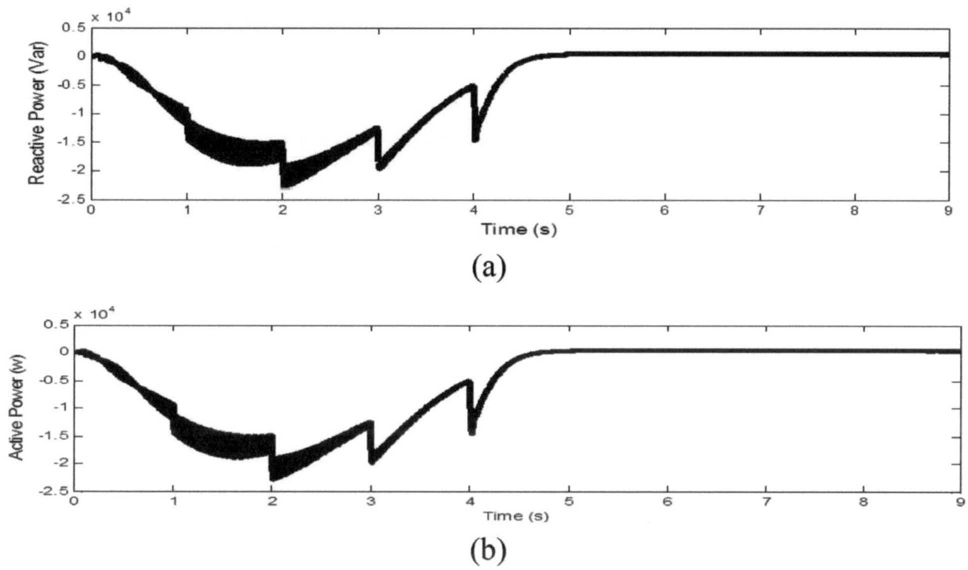

Figure 15.
(a) and (b). The output results of the active power (a) and reactive power (b), controlled using the FT1 controller.

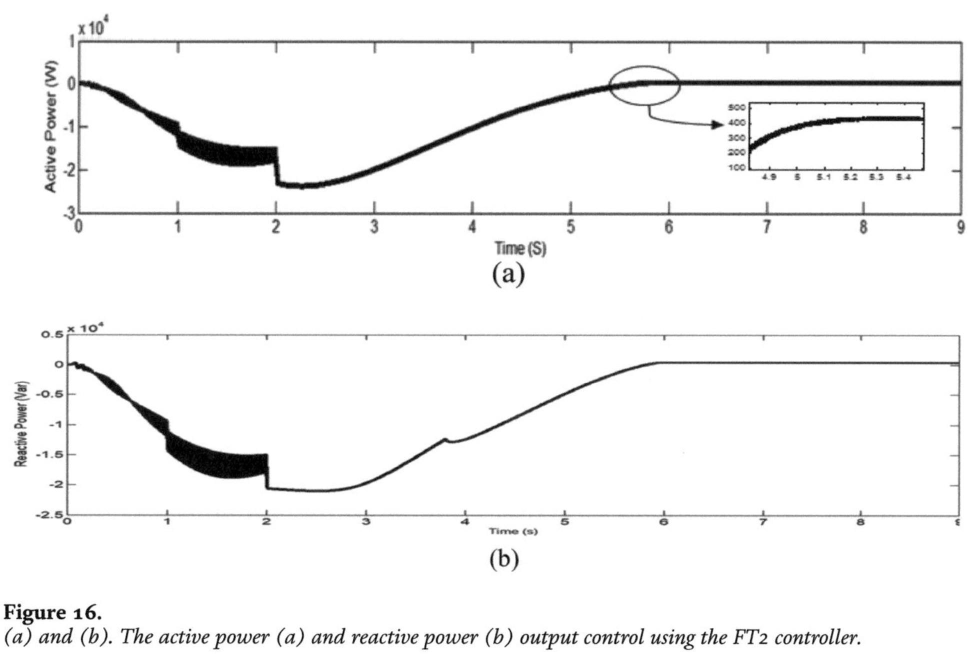

Figure 16.
(a) and (b). The active power (a) and reactive power (b) output control using the FT2 controller.

Parameters	Acronyms	Numerical values
Frequency	(F)	60
Stator line voltage	($V_{L(rms)}$)	200
Stator resistance	(R_S)	3.35
Rotor resistance	(R_r)	1.99
Stator leakage inductance	(L_{LS})	6.94e^{-3}
Rotor leakage inductance	(L_{Lrp})	6.94e^{-3}
Magnetic inductance	(L_M)	163.73e^{-3}
Moment inertia	(J)	0.1
Number of pole	(P)	4

Table 6.
Numerical values of the main parameters of the DFIG with its acronyms.

information, all the numerical values of the main parameters of the DFIG design with its acronyms are organized in **Table 6**. The DFIG numerical amount chart consists of voltage, resistance, inductance leakage of the rotor and stator, magnetic inductance, moment inertia, and the number of poles.

Notation 3: Since, in this paper, the main control aim is sustainable of the output powers without any fluctuation or is less overshoots at the reference value; hence, the numerical value 400 W is considered for both controllers, as the reference amount.

2.3.1 P & Q powers controlled using type-1 fuzzy controller

With regard to **Figure 15(a)** and **(b)**, both powers *Ps* and *Qs* of the type-1 controller's output, until reaching to the stable conditions, have faced multiple oscillation (overshoots) in its transient state, and also, by taking into consideration the presence of uncertainty, the powers are stabilized at the reference amount 400 W. The FT1 controller just covers a small interval in its output, and as the performance of the main system which is based on the uncertainty, the FT1 controller would not be able to properly control the output powers because of the presence of uncertainty, in which the transient states of FT1 controller output are with multiple fluctuations. However, multiple fluctuations occur in the FT1 controller's output mainly due to the uncertainty.

2.3.2 Ps & Q powers controlled using type-2 fuzzy controller

Figure 16 demonstrates the active and reactive powers outputs. Therefore, as shown, the performance of T2 controller is better than T1 controller; in other words, the powers have improved in its transient mood by considering the presence of the uncertainty, and on the one hand, T2FLC has a smoother surface in its control of the output powers. Since the computational burden of the mathematical theory of the type-2 fuzzy strategy is high, the output response of the FT2 controller until the stable state is associated with a time delay of several seconds. Due to the capability of the FT2 controller in covering a large range of the uncertainty, fluctuations have been removed, and it presents a smoother behavior in its transient state. In this paper, as previously described, with a little time delay, both P & Q power outputs of the FT1 and FT2 controller are stabilized at the value of 400 W. The stability of the active and reactive powers at the reference value has been depicted in **Figure 16(a)** and **(b)**, respectively. Ps and Q powers output response of T1FLC indicates the controlled output power stability at the pre-defined reference value in **Figure 13**. As specified in **Figure 16(a)**, it has been shown that the active power is stabilized at the value of 400 W. By taking into consideration the type-1 and type-2 controller functions, the major difference between them is to cover the uncertainty. The T1 controller cannot cover uncertainty, or in other words, it just controls the powers over a small specified range, while the T2 fuzzy controller can cover the uncertainty in large scales.

Both active power and reactive power output response of the type-1 FLC indicate the controlled output power stability at the pre-defined reference value in **Figure 16**. In **Figure 16(a)**, it has been shown that the active power is stabilized at the value of 400 W. Regarding the functions of type-1 and type-2 fuzzy controllers, the major difference between them is to cover the uncertainty. The T1 controller cannot cover uncertainty, or in other words, it just controls the powers over a small specified range, while the T2 fuzzy controller can cover the uncertainty in large scales.

2.3.3 Voltages [$U_{as}U_{bs}U_{cs}$] and currents [$I_{as}I_{bs}I_{cs}$] of the stator

In this section, the results of 3-phase voltage and currents between DFIG's stator and the main power grid through the power transmission lines have been investigated. Hence, the results of the stator 3-phase voltages are characterized in the frame of [$U_a U_b\ U_c$]. Also, in order to show the numerical range of sinusoidal output of the wind turbine, the numerical value 200 V is intended for the output system. **Figure 17(a)** indicates that the output voltage of the DFIG's stator is in the range of

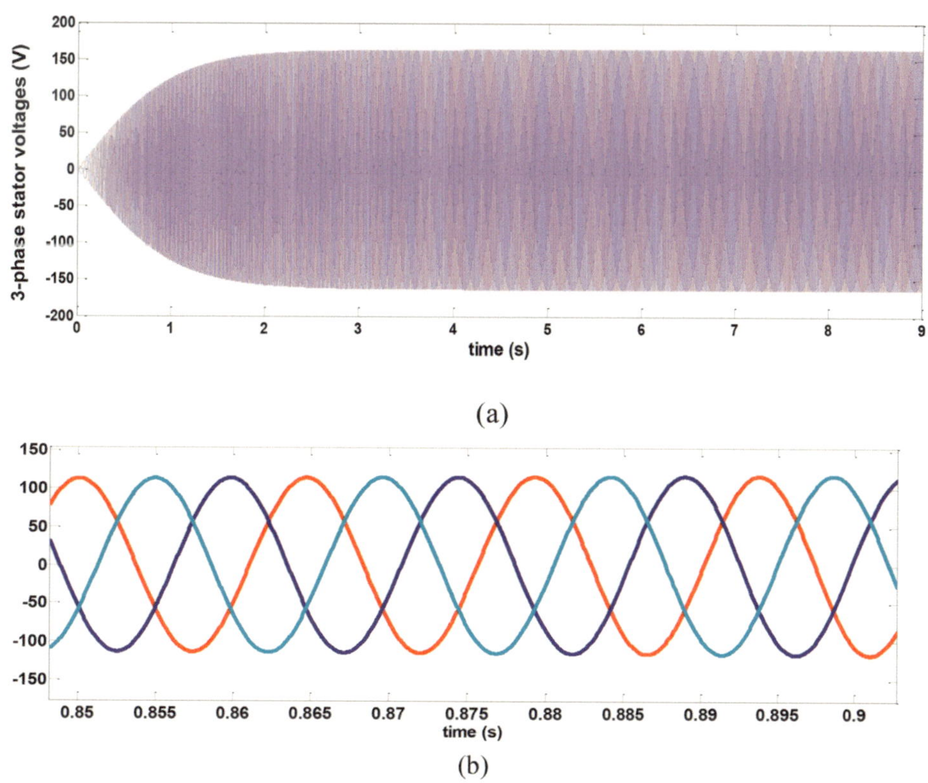

Figure 17.
(a) and (b). 3-phase voltages from grid to the stator in (a, b, c) reference frame with the sinusoidal waveform.

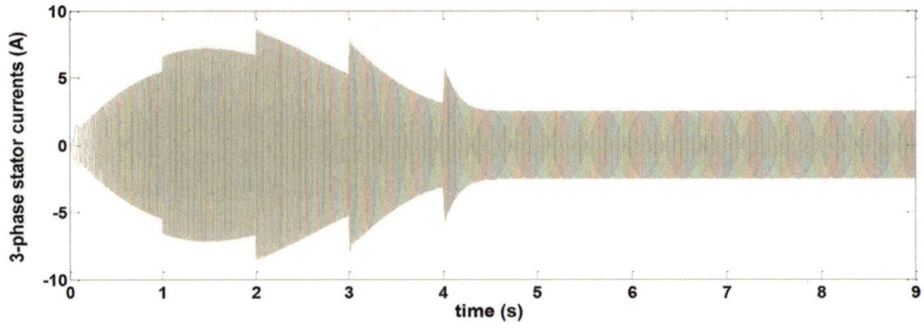

(a)

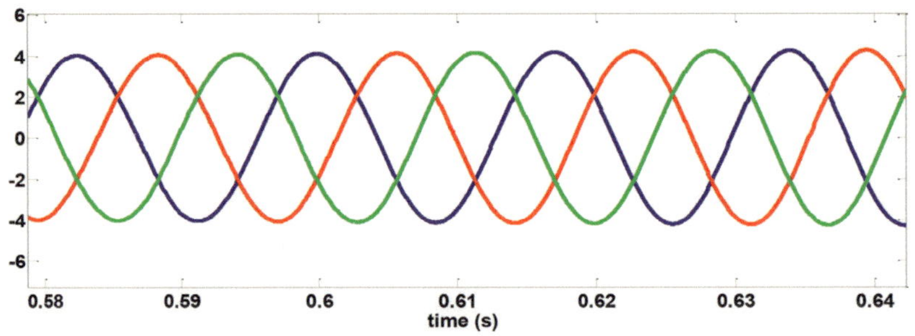

(b)

Figure 18.
(a) and (b). 3-phase currents of the stator in (a, b, c) reference frame with sinusoidal waveform.

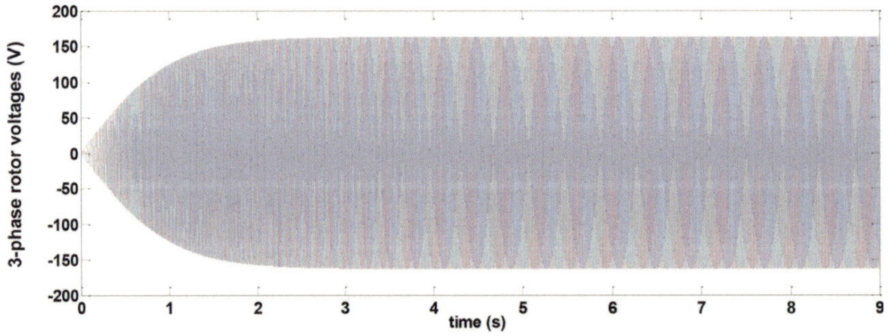

(a)

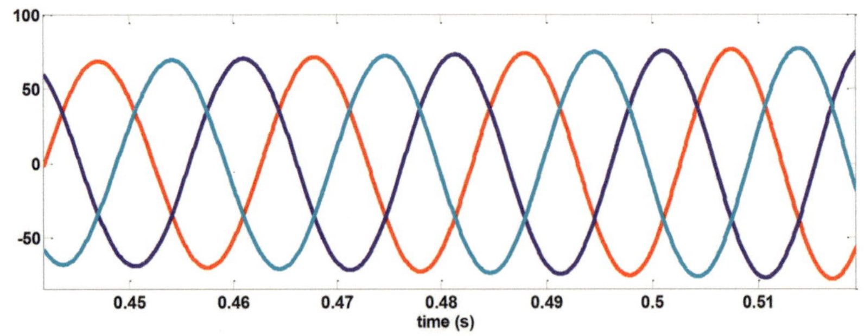

(b)

Figure 19.
(a) and (b). 3-phase rotor voltage with sinusoidal waveform.

(−200, +200) exactly. So, the sinusoidal waveform of the three-phase voltages has specified in **Figure 14(b)** clearly. **Figure 18(a)** is illustrates the stator current response by taking into consideration the range of (−10, +10). Generally, the waveforms of the stator 3-phase currents are sinusoidal as well, which have been depicted in **Figure 18(b)**.

2.3.4 Voltages [$U_{ar} U_{br} U_{cr}$] of the rotor

The input of the PWM is the voltage in [$a_r b_r c_r$] frame. Regarding the input voltages of the rotor, which will be in *a*, *b*, *c* frame, therefore, before the delivery of the signal from controller to the PWM, the output of the FT2 controller that is the voltage rotor in *d-q-0* frame should be converted to a, b, c reference system. The interval of the rotor's 3-phase voltage depends upon the considered value. In this paper, the desired numerical amount is (200 V). **Figure 19(a)** represents the three-phase rotor output voltage that is stable in the range of [−200, +200]. The value of the PWM unit input voltage must be in per unit. In principle, the rotor voltage waveform is sinusoidal which has been depicted in **Figure 19(b)**.

3. The possible structure of an offshore wind turbine

In general, the structure of ocean wind farms can be divided into two main categories:

1. From the perspective of the foundation used for construction

2. From the depth of water view

This chapter addresses the above issues.

3.1 Offshore wind turbine from foundation point of view

Offshore turbines are placed in the water and have more complexity to install on a turbine mounted on the land. **Figure 20** shows the different foundations for ocean wind turbines. Additionally, offshore wind turbine foundation must withstand harsh condition as well. This explains the wide variety of foundation developed over the years for offshore turbines, some more proven than others [54].

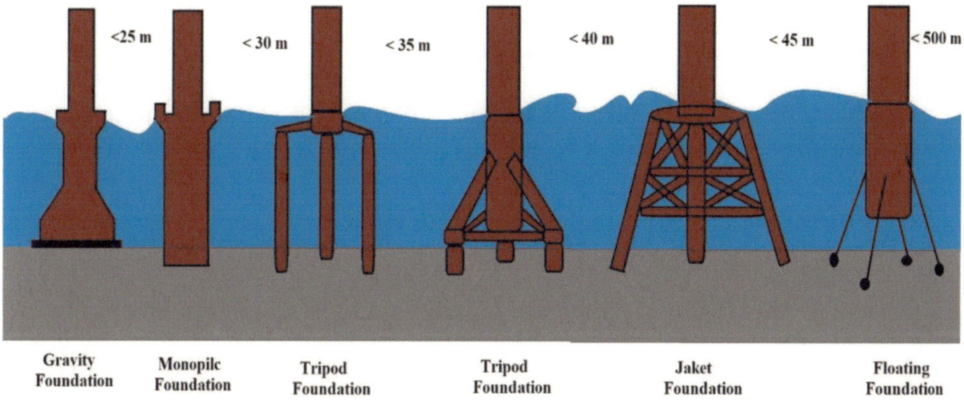

Figure 20.
The different foundations for ocean wind turbines.

3.2 Offshore wind turbine from the depth of water point of view

The layout of offshore wind farm changes based on the geographic area, the structure of the wind turbine, and the depth of water. The structure of water turbines in shallow water, deepwater, and floating has been investigated.

3.2.1 Shallow water offshore turbine

For areas with a water depth of fewer than 40 m, the use of offshore wind farms is appropriate.

Figure 21 shows the typical structure of shallow water wind turbines:

A. Gravity base

B. Mono-plie

C. Mono-caisson

D. Multi-pile

E. Multi-caisson

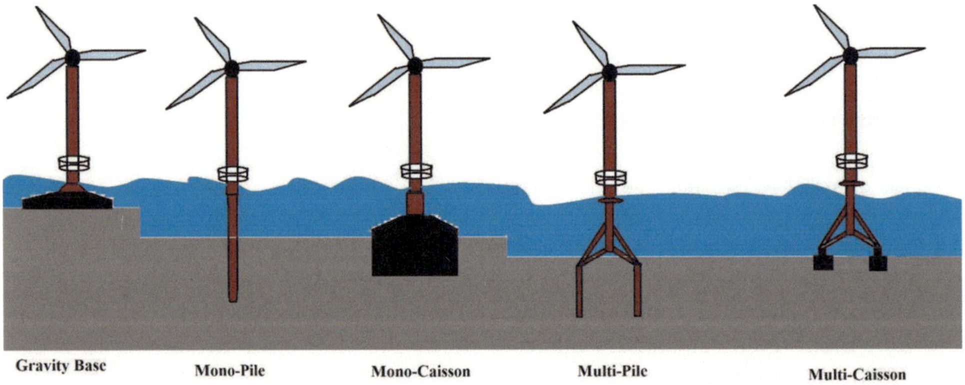

Figure 21.
Models of ocean wind turbines in shallow water.

3.2.2 Deepwater offshore wind turbine

For areas with a water depth of more than 40 m, the use of low wind turbine is appropriate. **Figure 22** shows the typical structure of deepwater wind turbines:

A. Tripod tube steel

B. Guyed tube

C. Spaceframe

D. Talisman energy concept

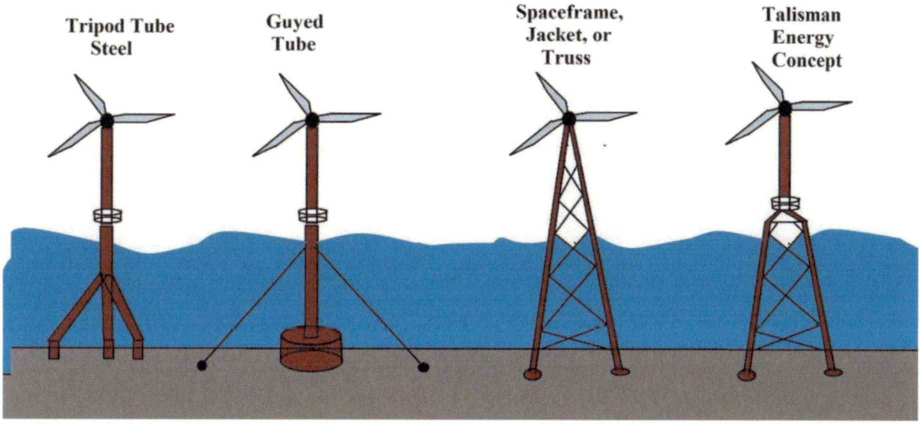

Figure 22.
Models of ocean wind turbines in deepwater.

3.2.3 Floating offshore wind turbine

Floating wind turbines are constructed on a floating structure on water, which is kept in different ways on the ocean floor. This method is used in areas where it is not possible to make a foundation for them. **Figure 23** shows three types of floating structures in an offshore wind turbine.

1. Tension leg mooring systems

2. Catenary mooring systems

3. Ballasted catenary configuration

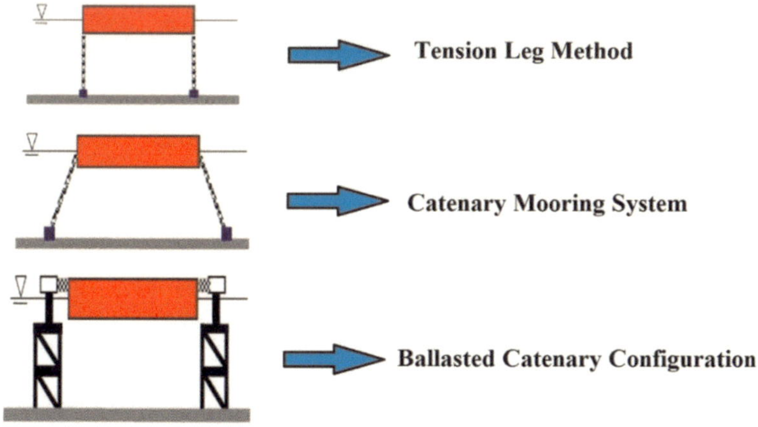

Figure 23.
Three types of engineered design for anchoring floating structures.

3.3 Offshore wind farm design

The design of offshore wind farms should be considered from three crucial points. **Figure 24** shows the design process for a typical ocean wind turbine.

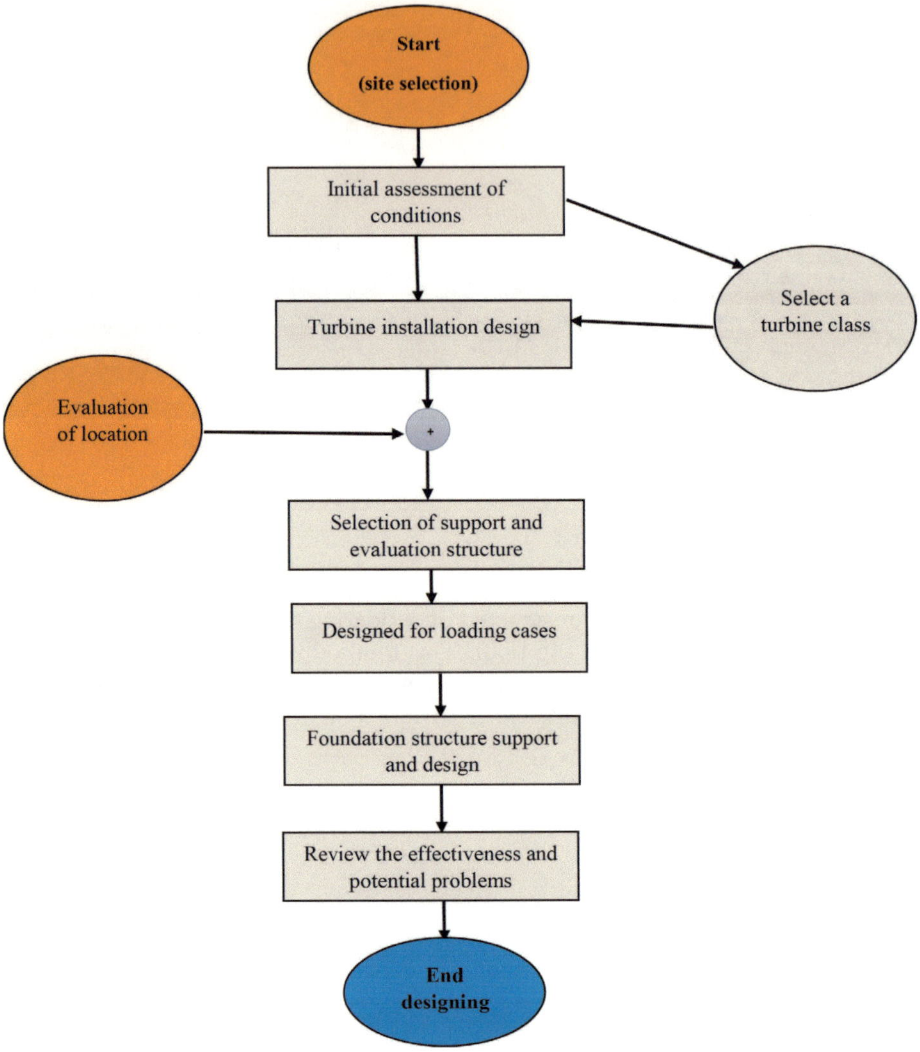

Figure 24.
The design process for a typical ocean wind turbine.

4. Challenges of offshore wind power plants

In general, two types of offshore wind power plant structures are challenging, (i) fixed offshore wind turbine and (ii) floating offshore wind turbine. Also, important issues that are being considered as the current challenge in offshore wind farms are the turbine layout and the way electricity is transmitted from the ocean to the shore [55].

Offshore installations currently consist of only a small percentage of the renewable energy market. However, due to the advancement of technology in the design and evaluation of these types of energy resources, it is expected that much progress will be made shortly. Offshore wind farms are in the early stages of their commercialization. They demand a higher cost of capital than onshore wind farms, but this can be compensated by higher capacity factors [56]. Offshore wind farms allow more widespread utilization of wind energy potentials. The reason for the higher capacity factors and the possibility of more use of offshore wind energy are as follows:

- There are no obstacles and restrictions for installing wind turbines for offshore turbines.
- It is possible to build sizeable enormous wind farms offshore.
- In places where the wind speed is average, it can be constructed.
- Offshore turbines have more extended and higher blades, leading to more swept areas and higher electricity outputs.

The offshore turbines are designed for use in offshore. Due to the lack of focus on such issues as shaking and impulse, noise, and visual contamination, there is a relatively different technical path. Although the issue of increasing turbine size for offshore turbines is a problem, this will increase the profit but also increase the operating costs. In this regard, the changing of the design and the ability to consider considerations are likely to provide better conditions in the design of offshore turbines.

Now, new turbines have a power of at least 5 MW. Therefore, a 1000 MW power plant can be achieved by installing 200 turbines [56]. Increase in the cost of the offshore wind turbine installation in the sea and the transfer of energy to the coast are most significant problems which need to be considered. Researchers are still trying to find the ways to reduce the cost of ocean wind farm.

Another challenge in the construction of offshore wind farms is the shortage of large ships that can carry large and heavy parts such as turbines. Also, another challenge in the field of offshore wind farms is the incentive to participate in the electricity market. Power transfer from the plant to the power grid using suitable infrastructures is also challenging for the use of this future energy. There are many other challenges that need to be addressed with the availability of sufficient technology in the world and companies who are active in wind turbine production [57]. Currently, the number of companies specialized in this field is insufficient, and it is expected that the number of these companies will increase shortly. Finally, to expand the use of this energy source, the training of a specialist who can build and

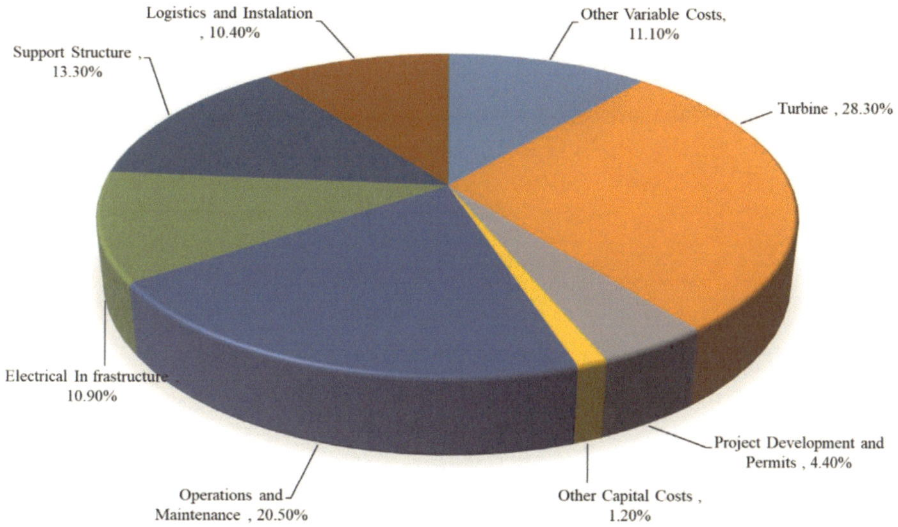

Figure 25.
The lifecycle of an offshore power plant.

operate offshore wind farms is another issue that should be addressed by electricity companies.

Regarding the abovementioned explanation, the major challenges of offshore wind technology are the high cost of offshore wind provision, lack of current infrastructure to support the fabrication such as installation, operation interconnection, maintenance of the system, and the challenges related to the lack of site data and lack of experience.

Figure 25 shows the lifecycle of an offshore power plant. According to this diagram, to solve the first challenge, it should be possible to reduce the impact of this problem in the long-term reports, with the development of industries and the reduction of installation costs and increased reliability of the system.

Nowadays, installation of ocean wind turbines requires specialized vessels, grid interconnections, purpose-built portside infrastructure, and robust undersea electricity transmission lines will be useful from the financial point of view. Regarding the last challenge, ocean wind projects confront new and untested allowing processes, which contributes to the uncertainty and risk faced by potential project developers and financiers.

Author details

Foad H. Gandoman[1,2], Abdollah Ahmadi[3], Shady H.E. Abdel Aleem[4*], Masoud Ardeshiri[5], Ali Esmaeel Nezhad[6], Joeri Van Mierlo[1,2] and Maitane Berecibar[1,2]

1 Research Group MOBI—Mobility, Logistics, and Automotive Technology Research Center, Vrije Universiteit Brussel, Brussels, Belgium

2 Flanders Make, Heverlee, Belgium

3 School of Electrical Engineering and Telecommunications, The University of New South Wales, Sydney, NSW, Australia

4 Mathematical, Physical and Engineering Sciences, 15th of May Higher Institute of Engineering, Cairo, Egypt

5 Department of Electrical Engineering, Kazerun Branch, Islamic Azad University, Kazerun, Iran

6 Department of Electrical, Electronic, and Information Engineering, University of Bologna, Italy

*Address all correspondence to: engyshady@ieee.org

IntechOpen

© 2020 The Author(s). Licensee IntechOpen. Distributed under the terms of the Creative Commons Attribution - NonCommercial 4.0 License (https://creativecommons.org/licenses/by-nc/4.0/), which permits use, distribution and reproduction for non-commercial purposes, provided the original is properly cited. (cc) BY-NC

References

[1] Blaabjerg F, Ma K. Wind energy systems. Proceedings of the IEEE. 2017; **105**(11):2116-2131

[2] Lua X, McElroya MB, Kiviluomac J. Global potential for wind-generated electricity. PNAS. 2009;**106**(27): 10933-10938

[3] Lua X, McElroya MB. Chapter 4— Global potential for wind-generated electricity. In: Wind Energy Engineering A Handbook for Onshore and Offshore Wind Turbines. Elsevier; 2017. pp. 51-73

[4] Zheng CW, Li CY, Pan J, Liu MY, Xia LL. An overview of global ocean wind energy resource evaluations. Renewable and Sustainable Energy Reviews. 2016;**53**:1240-1251

[5] Li W, Yao H, Wang H, Wang Z. Research progresses in assessment of China's offshore wind energy resources. Journal of Renewable and Sustainable Energy. 2014;**6**(05):31-38

[6] GWEC-Global-Wind-2015-Report_April-2016_22_04, www.gwec.net

[7] Leary D, Esteban M. Recent developments in offshore renewable energy in the Asia-Pacific region. Ocean Development & International Law. 2011;**42**:94-119

[8] Makridis C. Offshore wind power resource availability and prospects: A global approach. Environmental Science & Policy. 2014;**6**(05):31-38

[9] Mytilinou V, Kolios AJ, Di Lorenzo G. A comparative multi-disciplinary policy review in wind energy developments in Europe. International Journal of Sustainable Energy. 2015;**36**(8):754-774

[10] The European offshore wind industry. Key Trends and Statistics 2016. windeurope.org

[11] Belward A, Bisselink B, Bódis K, Brink A, Dallemand J-F, de Roo A, et al. Renewable energies in Africa. Luxembourg: Publications Office of the European Union; 2011. Available from: http://www.jrc.ec.europa.eu/

[12] Musial W, Heimiller D, Beiter P, Scott G, Draxl C. 2016 Offshore Wind Energy Resource Assessment for the United States. NERL, Denver West Parkway: National Renewable Energy Laboratory; 2016

[13] Daniel J, Liu S, Pan J. Potential impacts and economic value of US offshore wind. In: IEEE Power & Energy Society General Meeting; Denver, CO. 2015. pp. 1-5

[14] Sun X, Huang D, Wu G. The current state of offshore wind energy technology development. Energy. 2012; **41**(1):298-312

[15] Li W, Zhou H, Liu H, Lin Y, Xu Q. Review on the blade design technologies of tidal current turbine. Renewable and Sustainable Energy Reviews. 2016;**63**: 414-422

[16] Nejad AR, Guo Y, Gao Z, Moan1 T. Development of a 5 MW reference gearbox for offshore wind turbines. Wind Energy. 2016;**19**:1089-1106

[17] Toftevaag T. Specialization Course ELK 12. NTNU; 2012

[18] Blaabjerg F, Ma K. Future on power electronics for wind turbine systems. IEEE Journal of Emerging and Selected Topics in Power Electronics. 2013;**1**(3): 139-152

[19] Gandoman FH, Sharaf AM, Abdel Aleem SHE, Jurado F. Distributed FACTS stabilization scheme for efficient utilization of distributed wind energy systems. International Transactions

on Electrical Energy Systems. 2017; **27**(11):1-20

[20] Gandoman FH, Ahmadi A, Sharaf AM, Siano P, Pou J, Hredzak B, et al. Review of FACTS technologies and applications for power quality in smart grids with renewable energy systems. Renewable and Sustainable Energy Reviews. 2018;**82**:502-514

[21] El-mohr I. Renewable energy systems. Wind Energy. 2010. Available from: http://slideplayer.com/slide/10266678/

[22] Bhutto DK, Ansari JA, Chachar F, Katyara S, Soomro J. Selection of optimal controller for active and reactive power control of doubly fed induction generator (DFIG). In: Computing, Mathematics and Engineering Technologies (iCoMET), 2018 International Conference on. IEEE; 2018. pp. 1-5

[23] Wu C, Nian H, Pang B, Cheng P. Adaptive repetitive control of DFIG-DC system considering stator frequency variation. IEEE Transactions on Power Electronics. 2018;**34**(3):3302-3312

[24] Sáiz-Marín E, Lobato E, Egido I. New challenges to wind energy voltage control. Survey of recent practice and literature review. IET Renewable Power Generation. 2017;**12**(3):267-278

[25] Parida A, Chatterjee D. Integrated DFIG–SCIG-based wind energy conversion system equipped with improved power generation capability. IET Generation, Transmission & Distribution. 2017;**11**(15):3791-3800

[26] Lin F-J, Tan K-H, Tsai C-H. Improved differential evolution-based Elman neural network controller for squirrel-cage induction generator system. IET Renewable Power Generation. 2016;**10**(7):988-1001

[27] Wang Y, Meng J, Zhang X, Xu L. Control of PMSG-based wind turbines for system inertial response and power oscillation damping. IEEE Transactions on Sustainable Energy. 2015;**6**(2): 565-574

[28] Hu J, Nian H, Hu B, He Y, Zhu Z. Direct active and reactive power regulation of DFIG using sliding-mode control approach. IEEE Transactions on Energy Conversion. 2010;**25**(4): 1028-1039

[29] Naidu NS, Singh B. Grid-interfaced DFIG-based variable speed wind energy conversion system with power smoothening. IEEE Transactions on Sustainable Energy. 2017;**8**(1):51-58

[30] Kouadria S, Messlem Y, Berkouk EM. Sliding mode control of the active and reactive power of DFIG for variable-speed wind energy conversion system. In: Renewable and Sustainable Energy Conference (IRSEC), 2015 3rd International. Marrakech, Morocco: IEEE; 2015. pp. 1-8

[31] Ranga KP, Sukumar GD, Pakkiraiah B, Rao MS. Neuro fuzzy based P s &Q S controller for Doubly Fed Induction Generator with wind turbine. In: Electrical Power and Energy Systems (ICEPES), International Conference on. Bhopal, India: IEEE; 2016. pp. 139-144

[32] Mohanty A, Patra S, Ray PK. Robust fuzzy-sliding mode based UPFC controller for transient stability analysis in autonomous wind-diesel-PV hybrid system. IET Generation, Transmission & Distribution. 2016;**10**(5):1248-1257

[33] Hamane B, Doumbia M, Bouhamida M, Benghanem M. Control of wind turbine based on DFIG using Fuzzy-PI and Sliding Mode controllers. In: Ecological Vehicles and Renewable Energies (EVER), 2014 Ninth International Conference on. Monte-Carlo, Monaco: IEEE; 2014. pp. 1-8

[34] Li P, Wang J, Wu F, Li H. Nonlinear controller based on state feedback linearization for series-compensated DFIG-based wind power plants to mitigate sub-synchronous control interaction. International Transactions on Electrical Energy Systems. 2018;**29**: e2682

[35] Taher SA, Dehghani Arani Z, Rahimi M, Shahidehpour M. A new approach using combination of sliding mode control and feedback linearization for enhancing fault ride through capability of DFIG-based WT. International Transactions on Electrical Energy Systems. 2018;**28**:e2613

[36] Ashouri-Zadeh A, Toulabi M, Bahrami S, Ranjbar AM. Modification of DFIG's active power control loop for speed control enhancement and inertial frequency response. IEEE Transactions on Sustainable Energy. 2017;**8**(4): 1772-1782

[37] Ghosh S, Kamalasadan S. An integrated dynamic modeling and adaptive controller approach for flywheel augmented DFIG based wind system. IEEE Transactions on Power Systems. 2017;**32**(3):2161-2171

[38] Mendel JM, Wu D. Critique of "a new look at Type-2 fuzzy sets and Type-2 fuzzy logic systems". IEEE Transactions on Fuzzy Systems. 2017; **25**(3):725-727

[39] Raju SK, Pillai G. Design and implementation of type-2 fuzzy logic controller for DFIG-based wind energy systems in distribution networks. IEEE Transactions on Sustainable Energy. 2016;**7**(1):345-353

[40] Yassin HM, Hanafy HH, Hallouda MM. Enhancement low-voltage ride through capability of permanent magnet synchronous generator-based wind turbines using interval type-2 fuzzy control. IET Renewable Power Generation. 2016; **10**(3):339-348

[41] Khooban MH, Niknam T, Sha-Sadeghi M. Speed control of electrical vehicles: A time-varying proportional–integral controller-based type-2 fuzzy logic. IET Science, Measurement & Technology. 2016;**10**(3):185-192

[42] Sarabakha A, Fu C, Kayacan E, Kumbasar T. Type-2 fuzzy logic controllers made even simpler: From design to deployment for UAVs. IEEE Transactions on Industrial Electronics. 2018;**65**(6):5069-5077

[43] Chaoui H, Khayamy M, Aljarboua AA. Adaptive interval type-2 fuzzy logic control for PMSM drives with a modified reference frame. IEEE Transactions on Industrial Electronics. 2017;**64**(5):3786-3797

[44] Rahman NHA, Zobaa AF. Integrated mutation strategy with modified binary PSO algorithm for optimal PMUs placement. IEEE Transactions on Industrial Informatics. 2017;**13**(6):3124-3133

[45] Wei L-X, Li X, Fan R, Sun H, Hu Z-Y. A hybrid multiobjective particle swarm optimization algorithm based on R2 indicator. IEEE Access. 2018;**6**:14710-14721

[46] Mahdi FP, Vasant P, Abdullah-Al-Wadud M, Watada J, Kallimani V. A quantum-inspired particle swarm optimization approach for environmental/economic power dispatch problem using cubic criterion function. International Transactions on Electrical Energy Systems. 2018;**28**(3):e2497

[47] Serrano FE, Flores MA. C++ library for fuzzy type-2 controller design with particle swarm optimization tuning. In: Central American and Panama Convention (CONCAPAN XXXV), 2015 IEEE Thirty Fifth. Tegucigalpa, Honduras: IEEE; 2015. pp. 1-7

[48] Wang D, Ma N, Wei M, Liu Y. Parameters tuning of power system

stabilizer PSS4B using hybrid particle swarm optimization algorithm. International Transactions on Electrical Energy Systems. 2018;**28**:e2598

[49] Hurel J, Mandow A, García-Cerezo A. Tuning a fuzzy controller by particle swarm optimization for an active suspension system. In: IECON 2012-38th Annual Conference on IEEE Industrial Electronics Society. Montreal, QC, Canada: IEEE; 2012. pp. 2524-2529

[50] Lee JH, Song J-Y, Kim D-W, Kim J-W, Kim Y-J, Jung S-Y. Particle swarm optimization algorithm with intelligent particle number control for optimal design of electric machines. IEEE Transactions on Industrial Electronics. 2018;**65**(2):1791-1798

[51] Koad RB, Zobaa AF, El-Shahat A. A novel MPPT algorithm based on particle swarm optimization for photovoltaic systems. IEEE Transactions on Sustainable Energy. 2017;**8**(2):468-476

[52] Tulay G, İskender İ, Erdem H. Optimal tuning of a boost PFC converter PI controller using heuristic optimization methods. International Transactions on Electrical Energy Systems. 2017;**27**(12):e2458

[53] Javadi MS, Esmaeel Nezhad A. Intelligent particle swarm optimization augmented with chaotic searching technique to integrate distant energy resources. International Transactions on Electrical Energy Systems. 2017;**27**(12): e2447

[54] Ahmed SS. Finite element modeling of large diameter monopiles in dense sand for offshore wind turbine foundation. In: Proceedings of the ASME 2015 34th International Conference on Ocean, Offshore and Arctic Engineering, May 31–June 5, 2015, Newfoundland, Canada

[55] Malhotra S. "Design and Construction Considerations for Offshore Wind Turbine Foundations", Research and Innovations. New York: P D Network; 2004

[56] Oh KY, Nam W, Ryu MS, Kim J, Epureanu BI. A review of foundations of offshore wind energy convertors: Current status and future perspectives. Renewable and Sustainable Energy Reviews. 2018;**88**:16-36

[57] Santos AC, Perez JP, Diez DB, Rodríguez C. Offshore wind energy: A review of the current status, challenges and future development in Spain. Renewable and Sustainable Energy Reviews. 2016;**64**:1-18

Chapter 5

An Adaptive Load Frequency Control Based on Least Square Method

Adelhard Beni Rehiara, Naoto Yorino, Yutaka Sasaki and Yoshifumi Zoka

Abstract

Modern power system becomes a complex system consisted with various load and power stations. Therefore, it may spread into some areas of power system in neighborhood, and so a load frequency control (LFC) is a necessity device to regulate the frequency of the power system by distributing the load to the generating units and controlling tie-line power interchange between areas. The integration of renewable energy sources (RES) into a power grid has presented important issues about stability and security of power system. In such conditions, the use of conventional LFC may not be sufficient to protect the power system against the power changes. In this chapter, an adaptive LFC controller is proposed using the least square method (LSM). The controller adopts an internal model control (IMC) structure in two scenarios, i.e., static controller gain with adaptive internal model and both the adaptive controller gain and adaptive internal model. A two-area power system is used to test and to validate both performance and the effectiveness of this controller through some case studies.

Keywords: adaptive controller, power system stability, LSM, LFC, IMC, res

1. Introduction

Demand load in a power system is continuously varied by the time and the change of the active and reactive power demand has introduced generator-load mismatching power. When the load increases, it will slow down the rotor speed which result drop of the frequency. On the other hand, when the load is reduced, the load frequency will rise. The change of load frequency will directly affect electrical motor performance, further interfering the protection of the system [1].

Smart grid technology is well developed, and it currently has been widely used in the power system operation due to the need for integrating renewable energy to the existing power grid. Penetration of the renewable energy resources such as solar generations and wind turbines to the power grid has introduced significant issues of stability and security of the power system. However, frequency stability is a major issue in the power system operation due to continuous output change of the renewables. Therefore a load frequency control (LFC) is an essential device to back up the automatic generation control (AGC) to keep the frequency stable during power system operation [2].

The frequency controllers in a power system consist of AGC and LFC systems as primary control and secondary control. An AGC will respond to small changes of frequency in a generator, and an LFC will have to regulate load frequency in a large area of the power system and large change in load frequency which is typically in between ±5 and 6% of the frequency bias [2–5].

In this chapter, adaptive LFC controllers are introduced, where an adaptive IMC model that is repeatedly updated by the least square method (LSM) in real-time operation is proposed. It is shown that the target model is successfully identified and therefore that the proposed LFC controller scheme effectively keeps the system frequency at a desired set point. The effectiveness of the proposed controller is demonstrated by the simulations using a standard LFC model representing two-area interconnected power system.

2. Power system model

2.1 Mathematical modelling of generator

Eq. (1) is well known as a swing equation that describes the rotor dynamics and hence is known as the swing Equation [6, 7]. The internal EMF angle δ is called the load angle that indicates how much power can be transferred:

$$\frac{2H}{\omega_s}\frac{d^2\delta}{dt^2} = P_m - P_e \tag{1}$$

For small perturbation, the swing equation of a synchronous machine will be formed in Eqs. (2) and (3) for its small deviation in speed. Therefore, Laplace transformation of Eq. (3) is shown in Eqs. (4) [6, 7]:

$$\frac{2H}{\omega}\frac{d^2\Delta\delta}{dt^2} = \Delta P_m - \Delta P_e \tag{2}$$

$$\frac{d\Delta\frac{\omega}{\omega_s}}{dt} = \frac{1}{2H}(\Delta P_m - \Delta P_e) \tag{3}$$

$$\Delta f = \frac{1}{2Hs}(\Delta P_m - \Delta P_e) \tag{4}$$

The mathematical model of the generator as formulated in Eq. (4) can be figured in **Figure 1**.

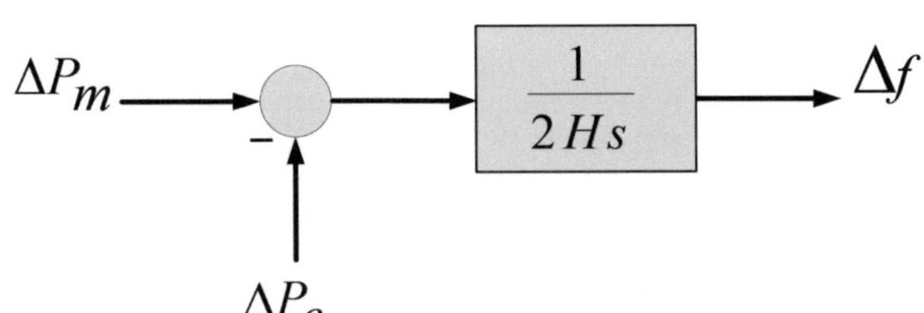

Figure 1.
Block diagram for generator.

2.2 Mathematical modelling of load

The load on a power system consists of a variety of electrical devices which is a resistive or inductive load. The equipment used for lighting purposes and heating are basically resistive in nature, and this kind of load is independent to frequency. On the other hand, rotating devices, such as fans and pumps, are basically a composite of the resistive and inductive components which are dependent to the frequency changes. The speed-load characteristic of the composite load is given by [6, 7]:

$$\Delta P_e = \Delta P_L + D\Delta f \tag{5}$$

where ΔP_L is the load change and $D\Delta f$ is the frequency sensitive load change. D is expressed as percent change in load by percent change in frequency. By adding the load to the generator, block diagram of both load and generator is figured out in **Figure 2**.

2.3 Mathematical modelling for prime mover

The electrical energy is generated inside a power generation by converting the other kind of energy sources by means of a prime mover. The prime mover may be diesel machines, hydraulic turbines at waterfalls, or steam turbines. The model for the turbine relates the changes in mechanical power output ΔP_m to the changes in the steam valve position ΔP_V [6, 7]:

$$G_T = \frac{\Delta P_m}{\Delta P_v} = \frac{1}{1 + T_t s} \tag{6}$$

Figure 3 expresses the prime mover block diagram, where T_t is the turbine constant which has the range in between 0.2 and 2.0 seconds.

2.4 Mathematical modelling for governor

The electrical power will exceed the mechanical power input when the electrical load is suddenly increased. This condition will result in the extraction from the

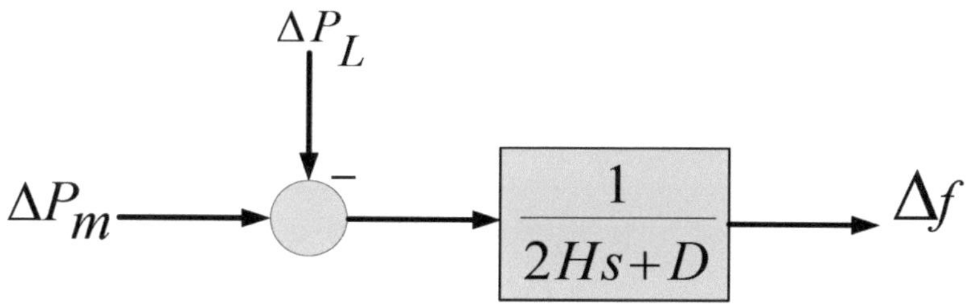

Figure 2.
Generator and load block diagram.

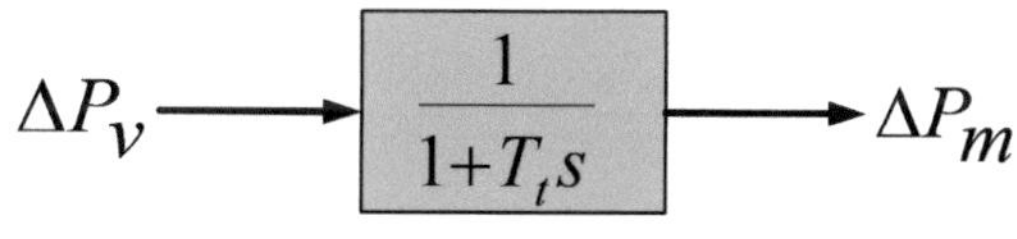

Figure 3.
Block diagram for prime mover.

rotating energy of the turbine. Then the kinetic energy stored in the machine is reduced and slows down the speed of prime mover. Therefore, to compensate the reduced speed, the governor sends a command to supply more volumes of water or steam or gas to increase the prime mover speed.

Speed regulation R is given as the curve slope in **Figure 4**. The typical speed regulation values of generator are in between 5 and 6% from zero to maximum of load [6, 7]:

$$\Delta P_g = \Delta P_{ref} - \frac{1}{R}\Delta f \tag{7}$$

The quantity of ΔP_g is converted to the position of steam valve ΔP_V by a governor time constant T_g. Therefore, the s-domain relation of ΔP_V and ΔP_g is a linear relationship by considering the simple time constant T_g [6, 7] (**Figure 5**):

$$\Delta P_v = \frac{1}{1 + T_g s}\,\Delta P_g \frac{1}{1 + T_g s} \tag{8}$$

Finally, **Figure 6** summarizes a combining of all of the block diagrams from earlier block diagrams for a single area system.

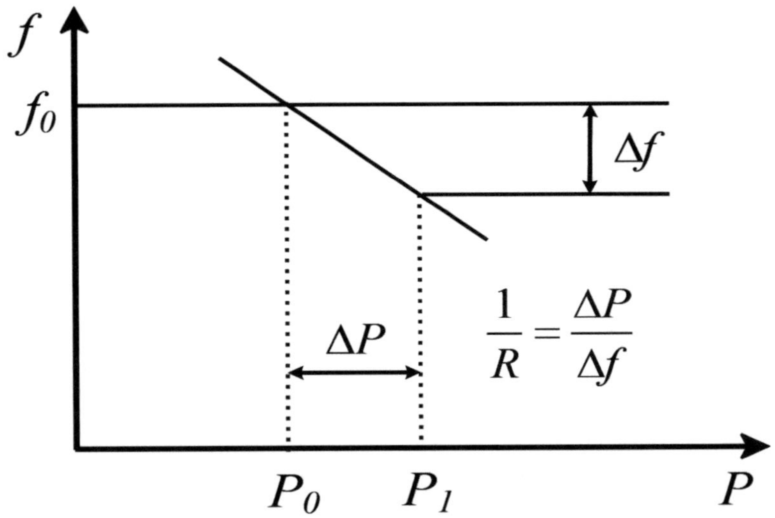

Figure 4.
Speed drop regulation.

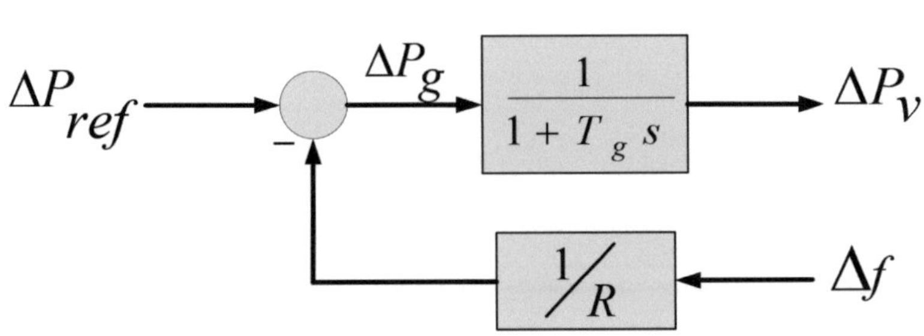

Figure 5.
Block diagram for governor.

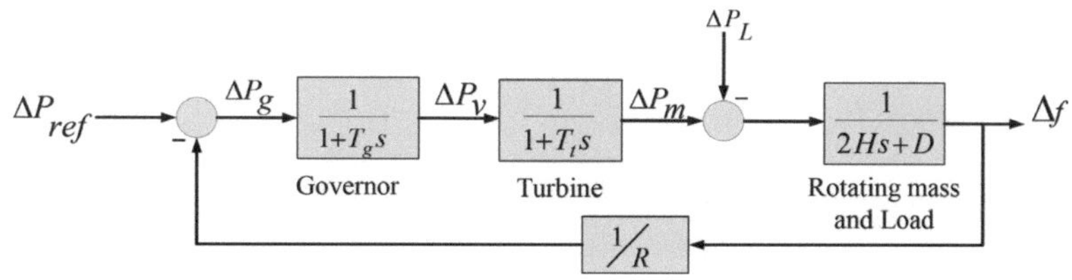

Figure 6.
Completed power system block diagram.

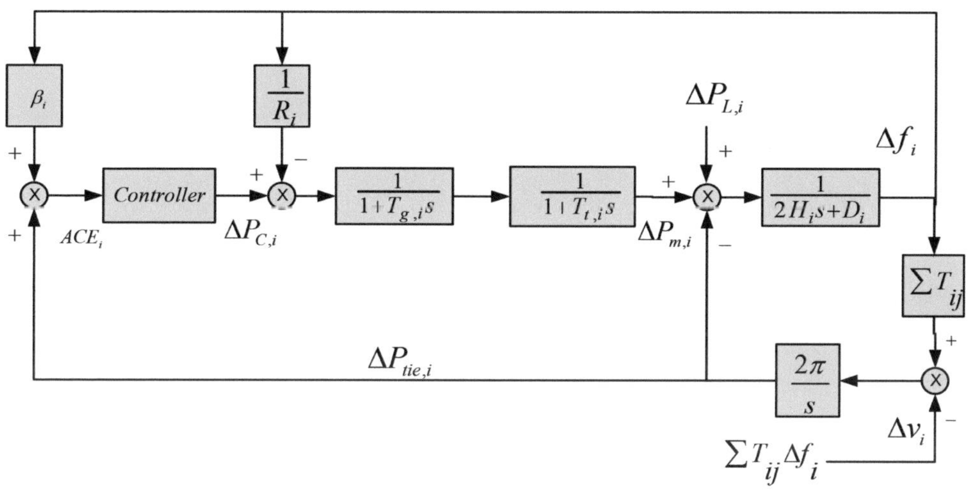

Figure 7.
Power system dynamics.

A completed LFC block diagram for multi-area power system, including controller, frequency bias, and tie-line power change, can be redrawn in **Figure 7** [2–4, 8].

The tie-line power change P_{tie} is calculated for all area n using Eq. (9), and the area control error (ACE) which is a suitable linear combination of frequency f and tie-line power changes for each area is found using Eq. (10) as follows [3, 4, 9]:

$$\Delta P_{tie,i} = \frac{2\pi}{s}\left[\sum_{\substack{j=1 \\ j\neq i}}^{n} T_{ij}\Delta f_i - \sum_{\substack{j=1 \\ j\neq i}}^{n} T_{ij}\Delta f_j\right] \tag{9}$$

$$ACE_i = \Delta P_{tie,i} + \beta_i \Delta f_i \tag{10}$$

A general state-space model is used to describe the power system model as shown in Eqs. (11) and (12):

$$\dot{x}(t) = Ax(t) + Bu(t) + Fw(t) \tag{11}$$

$$y(t) = Cx(t) + Du(t) \tag{12}$$

where $x(t)$, $u(t)$, $w(t)$, and $y(t)$ are the matrices of state variables, input variables, control variable, and output variable, respectively. Four variables of the state variables are $\Delta P_{g,i}$, $\Delta P_{m,i}$, Δf_i, and $\Delta P_{tie,I}$, and the input variables are $\Delta P_{L,i}$ and Δv_i. $\Delta P_{c,I}$ is the control variable, while ACE_i is the output variable.

Due to no direct connection between input and output variables, the feed forward matrix D is removed from the model. Therefore the system matrices of a LFC system are written in Eqs. (13)–(16) [3, 4, 9]:

$$A_i = \begin{bmatrix} -\dfrac{1}{T_{g,i}} & 0 & -\dfrac{1}{R_i T_{g,i}} & 0 \\ \dfrac{1}{T_{t,i}} & -\dfrac{1}{T_{t,i}} & 0 & 0 \\ 0 & \dfrac{1}{2H_i} & -\dfrac{D_i}{2H_i} & -\dfrac{1}{2H_i} \\ 0 & 0 & 2\pi \sum\limits_{\substack{j=1 \\ j \neq i}}^{N} T_{ij} & 0 \end{bmatrix} \tag{13}$$

$$B_i = \begin{bmatrix} \dfrac{1}{T_{g,i}} & 0 & 0 & 0 \end{bmatrix}^T \tag{14}$$

$$C_i = \begin{bmatrix} 0 & 0 & \beta_i & 1 \end{bmatrix} \tag{15}$$

$$F_i = \begin{bmatrix} 0 & 0 \\ 0 & 0 \\ -\dfrac{1}{2H_i} & 0 \\ 0 & -2\pi \end{bmatrix} \tag{16}$$

where $P_{g,i}$, $P_{m,i}$, $P_{L,i}$, $P_{c,i}$, y_i, H_i, d_i, R_i, β_i. T_{ij}, $T_{g,i}$, and $T_{t,i}$ are the output of governor, the prime mover power, the load, the control action, the output of system, the inertia constant, the damping coefficient, the characteristic of speed droop, the bias factor of frequency, the tie-line synchronizing coefficient between reference area i and area j, and the time constant of governor and turbine, respectively.

2.5 System response of power change

Consider a single machine system connected to an infinite bus as shown in **Figure 8**, and its swing equation in steady-state condition can be expressed in Eq. (17) [6]:

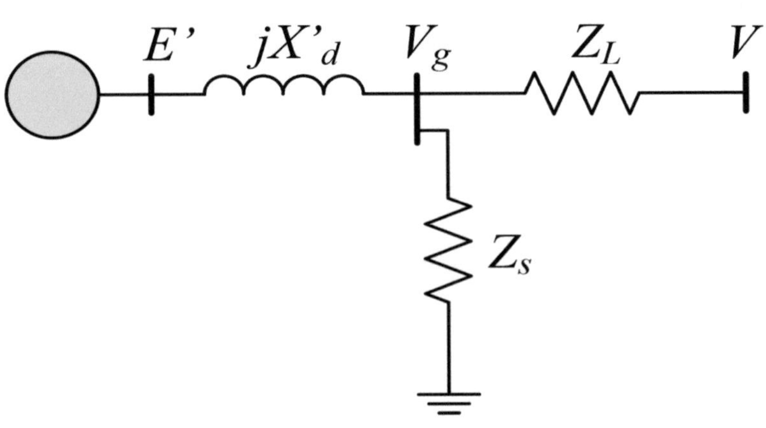

Figure 8.
Single machine system.

$$\frac{H}{\pi f_0}\frac{d^2\delta}{dt^2} = \Delta P_m - \Delta P_{\max}\sin\delta \tag{17}$$

If there is some change in mechanical power input P_m as the result of disturbances or load changes, power angle δ will change to a new state as $\delta = \delta 0 + \Delta\delta$. Then it will further influence swing equation into Eq. (18). The change affects the swing equation in terms of incremental changes in power angle as in Eq. (19):

$$\frac{H}{\pi f_0}\frac{d^2\delta_0}{dt^2} + \frac{H}{\pi f_0}\frac{d^2\Delta\delta}{dt^2} = \Delta P_m - \Delta P_{\max}\sin\delta_0 - \Delta P_{\max}\cos\delta_0\Delta\delta \tag{18}$$

$$\frac{H}{\pi f_0}\frac{d^2\Delta\delta}{dt^2} + \Delta P_{\max}\cos\delta_0\Delta\delta = 0 \tag{19}$$

The quantity of $P_{max}\ cos\ \delta_0$ is known as the synchronizing coefficient P_s which is the slope of power angle curve at δ_0. The root(s) of the second-order differential equation in Eq. (19) can be shown in Eq. (20):

$$s^2 = -\frac{\pi f_0}{H} P_S \tag{20}$$

It can be seen from Eq. (20) that there are possibilities of roots in s-plane when P_s is either positive or negative. A root in the right hand s-plane, where causes system unstable and responses increased exponentially, is gotten when synchronizing coefficient P_s is negative. In other way, two roots will be on j-ω axes of s-plane for P_s negative that causes system responses, oscillatory and undamped with natural frequency as in Eq. (21):

$$\omega_n = \sqrt{\frac{\pi f_0}{H} P_S} \tag{21}$$

3. Controller structure

3.1 Model predictive control

Model predictive control (MPC) is an advance optimal control in the field of control systems engineering. In an MPC, the optimal trajectory movement is given by properly choosing the MPC gain so that the control errors can be minimized.

The objective of the predictive control is to compute the manipulated variable u in order to optimize the output behavior of a controlled plant y. An MPC will use its internal model to calculate the manipulated variable [10] (**Figure 9**).

At a given discrete time k, the plant output is estimated through prediction horizon Np from time $k + 1$ to $k + Np$, and the MPC controller output is predicted by control horizon Nc. The output of the plant is continued to be minimized based on specified objective function which is typically in the form of a quadratic function as shown in Eq. (22) [10]:

$$J(k) = \sum_{m=1}^{Np} \Delta\hat{y}(k+m|k)^{\mathrm{T}} Q\ \Delta\hat{y}(k+m|k) + \sum_{m=1}^{Nc} \Delta u(k+m|k)^{\mathrm{T}} R\ \Delta u(k+m|k) + \dots$$

$$\dots + \sum_{m=1}^{Np} u(k+m|k)^{\mathrm{T}} N\ u(k+m|k) \tag{22}$$

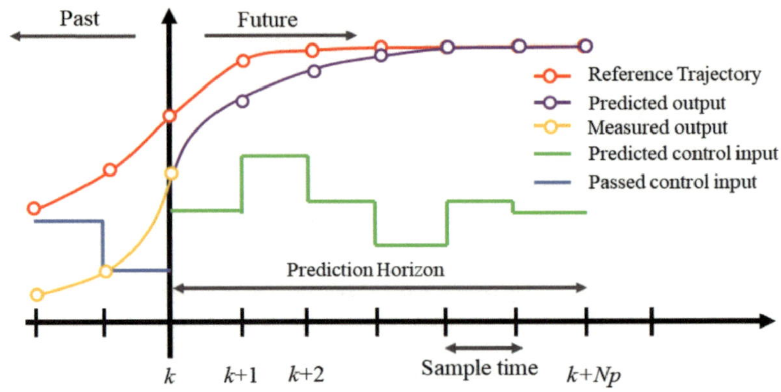

Figure 9.
Principle of MPC.

where N, Q, and R are the error weight matrices of the control action, output system, and rate of change in control action, respectively. $\Delta\hat{y}(k+m|k)$ is the output of plant in prediction, $\Delta u(k+m|k)$ is the change rate in control action under prediction, and $u(k+m|k)$ is the prediction of the action of optimal control, for the measurements given at time $k+m$ from the reference time k.

In order to have optimal result, prediction horizon and control horizon have to be set properly so that the MPC controller can work in high performance. On the other hand, absence of doing this will cause the MPC to lose optimal action that will result in a high overshoot response.

In a short time control horizon, *Nc* will respond with high control action that results in an overshoot after the end of control horizon time. While long time control horizon bring the controller be aggressive and used much energy to accelerate and decelerate the control action in order to keep it constant at set point. In same way, too long prediction horizon *Np* will drop controller performance due to the extra time needed for calculating the trajectory movement, while short time prediction horizon will cause inaccurate calculation of the trajectory.

For a given a linear system in continues time, a system state condition is expressed in Eq. (23) while Eq. (24) shows its formulation for discrete system with sampling time k.

$$\dot{x} = Ax + Bu \tag{23}$$

$$x(k+1) = Ax(k) + Bu(k) \tag{24}$$

$$y(k) = Cx(k) \tag{25}$$

Then the model will be converted to an augmented model so that later the quadratic programming problem with respect to ΔU could be formed easily. Because $u(k) = u(k\text{-}1) + \Delta u(k)$, then Eq. (24) can be rewritten as in Eq. (26), and its state space from is given in Eqs. (27) and (28) [11]:

$$x(k+1) = Ax(k) + Bu(k-1) + B\Delta u(k) \tag{26}$$

$$\begin{bmatrix} x(k+1) \\ u(k) \end{bmatrix} = \begin{bmatrix} A & B \\ 0 & 1 \end{bmatrix} \begin{bmatrix} x(k) \\ u(k-1) \end{bmatrix} + \begin{bmatrix} B \\ 1 \end{bmatrix} \Delta u(k) \tag{27}$$

$$y(k) = \begin{bmatrix} C & 0 \end{bmatrix} \begin{bmatrix} x(k) \\ u(k-1) \end{bmatrix} \tag{28}$$

For the *Np* prediction horizon based on the above equation model, the output of the prediction could be written as:

$$\begin{bmatrix} \hat{y}(k+1) \\ \hat{y}(k+2) \\ \vdots \\ \hat{y}(k+Np) \end{bmatrix} = \begin{bmatrix} CA \\ CA^2 \\ \vdots \\ CA^{Np} \end{bmatrix} x(k) + \begin{bmatrix} CB \\ CAB+CB \\ \vdots \\ \sum_{i=0}^{Np-1} CAB \end{bmatrix} u(k-1) + \dots \cdots$$
$$+ \begin{bmatrix} CB & 0 & \cdots & 0 \\ CAB+CB & CB & \cdots & 0 \\ \cdots & \cdots & \cdots & \cdots \\ \sum_{i=0}^{Np-1} CAB & \sum_{i=0}^{Np-2} CAB & \cdots & CB \end{bmatrix} \begin{bmatrix} \Delta u(k+1) \\ \Delta u(k+2) \\ \vdots \\ \Delta u(k+Np) \end{bmatrix} \quad (29)$$

Eq. (29) can be simplified as in Eq. (30):

$$\hat{Y} = \Omega X + \Pi u + G\Delta U = F + G\Delta U \quad (30)$$

Then the objective function could be written as:

$$\begin{aligned} J &= \left(\hat{Y} - r\right) + {}^{\mathrm{T}}\left(\hat{Y} - r\right) + \lambda \Delta U^{\mathrm{T}} \Delta U \\ &= (F + G\Delta U - r)^{\mathrm{T}} (F + G\Delta U - r) + \lambda \Delta U^{\mathrm{T}} \Delta U \\ &= \Delta U^{\mathrm{T}} \left(G^{\mathrm{T}} G + \lambda I\right) \Delta U + 2\Delta U^{\mathrm{T}} G^{\mathrm{T}} (F - r) \end{aligned} \quad (31)$$

Therefore, Eq. (31) could be solved efficiently by a quadratic programming.

3.2 Internal model control (IMC)

An IMC controller structure can be seen on **Figure 10** where an internal model is used parallel with the plant. This internal model, which is also called efferent model, used plant model to estimate plant output. It is known that to find the internal model as same as plant model is difficult due to plant dynamics that is not well captured in the modeling part.

In an IMC controller, an efferent model is expected to correct the actual output before it is feedback to the controller. This is the difference between MPC controllers with classical controllers. In order to get maximum performance, the efferent

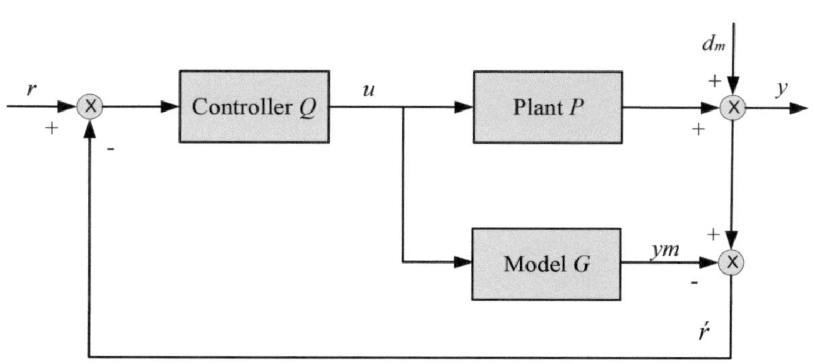

Figure 10.
IMC structure.

model may have input and output relation as close to the real plant model so that the only feedback signal is the disturbance.

As a type of controller, an IMC can control a plant directly. This controller can also be used to tune another controller if it is difficult to be the best parameter for this controller. Based on the IMC structure, this main controller can be any type of controller so that this IMC can be combined with any type of controllers [12]. The control law for an IMC control is written in Eq. (32) to Eq. (34) [13, 14]:

$$y = PQr + (1 - GQ)d \tag{32}$$

$$u = Qr - Qd \tag{33}$$

$$e = (1 - PQ)r - (1 - GQ)d \tag{34}$$

The difference between classical controller and an IMC is that an IMC will correct the actual output before it is fed back. Since an IMC uses the efferent model, the model should be a perfect model to have the highest control performance. The way to provide the model in an IMC can be in forward model, inverse models, combination of both forward and inverse models, or adaptive model.

3.3 Adaptive model structure

Due to highly nonlinearity in a power system, the classical model of power system may not be accurate to configure the real power system. Therefore the other objective of this paper is to find simple model of a load frequency of power system using the least square method (LSM). In order to capture the essential dynamics of power system, a first-order lag model is adopted in this chapter as follows:

$$G(s) = \frac{K}{1 + sT} \tag{35}$$

The power system model, including governor, turbine, and rotating mass and load model as in **Figure 7** will be replaced by the first-order lag model in Eq. (35). Therefore it has become a simple model as formulated in Eq. (36):

$$\Delta f_i = G_i(s) P_{C,i} \tag{36}$$

Expanding Eq. (35) gives

$$\Delta \dot{x}_i = A_i x_i + B_i u_i \tag{37}$$

$$\Delta \dot{f}_i = -\frac{1}{T_i}\Delta f_i + \frac{K_i}{T_i}\Delta P_{C,i} = A_i \Delta f_i + B_i \Delta P_{C,i} \tag{38}$$

For discrete time system, the above equation can be described in a matrix form as follows:

$$\begin{bmatrix} \Delta f_i(k+1) \\ \Delta f_i(k+2) \\ \vdots \\ \Delta f_i(k+M) \end{bmatrix} = [A_i] \begin{bmatrix} \Delta f_i(k) \\ \Delta f_i(k+1) \\ \vdots \\ \Delta f_i(k+M-1) \end{bmatrix} + [B_i] \begin{bmatrix} \Delta P_{C,i}(k) \\ \Delta P_{C,i}(k+1) \\ \vdots \\ \Delta P_{C,i}(k+M-1) \end{bmatrix} \tag{39}$$

Consider a linear equation of least square method $y = Hx$; a solution for minimizing the error can be written as [15, 16]:

$$J(x) = \|y - Hx\|^2 \tag{40}$$

Then expanding $J(x)$ gives

$$\begin{aligned} J(x) &= (y - Hx)^{\mathrm{T}}(y - Hx) \\ &= y^{\mathrm{T}}y - y^{\mathrm{T}}Hx - x^{\mathrm{T}}H^{\mathrm{T}}y + x^{\mathrm{T}}H^{\mathrm{T}}Hx \\ &= y^{\mathrm{T}}y - 2y^{\mathrm{T}}Hx + x^{\mathrm{T}}H^{\mathrm{T}}Hx \end{aligned} \tag{41}$$

Taking the derivative for the $J(x)$ gives

$$\frac{\partial}{\partial x}J(x) = -2H^{\mathrm{T}}y + 2H^{\mathrm{T}}Hx \tag{42}$$

Minimizing the derivative by setting it to zero gives

$$H^{\mathrm{T}}Hx = H^{\mathrm{T}}y \tag{43}$$

If $H^{\mathrm{T}}H$ is invertible, then the least square solution is given in Eqs. (44) and (45) for its simple form:

$$x = \left(H^{\mathrm{T}}H\right)^{-1}H^{\mathrm{T}}y \tag{44}$$

$$x = H^{\dagger}y \tag{45}$$

The least square solution for the discrete time system of the power system model is shown in Eq. (46).

$$\begin{bmatrix} \Delta f_{i,k+1} \\ \\ \vdots \\ \Delta f_{i,k+M} \end{bmatrix} = \begin{bmatrix} \Delta f_{1,k} & \cdots & \Delta f_{n,k} & \Delta P_{C1,k} & \cdots & \Delta P_{Cm,k} \\ \vdots & & \vdots & \vdots & & \vdots \\ \vdots & & \vdots & \vdots & & \vdots \\ \Delta f_{1,k+M-1} & \cdots & \Delta f_{n,k+M-1} & \Delta P_{C1,k+M-1} & \cdots & \Delta P_{Cm,k+M-1} \end{bmatrix} + \begin{bmatrix} A_{i,1} \\ \vdots \\ A_{i,n} \\ B_{i,1} \\ \vdots \\ B_{i,m} \end{bmatrix} \tag{46}$$

where $H^{\dagger}$ is the pseudo inverse of the H matrix. A and B matrices are taken from the LFC state-space model, and ACE is selected as input u to the LFC controller.

3.4 Model simplification

In order to adopt the least square solution into the internal model of the LFC system, the complex system of the LFC is likely to be reduced into a first-order model. After the disturbance is entering the LFC system, generations are expected to change as fast as possible to meet the load demand. This expectation may not be achieved due to slow response of governor and turbine operations. For these purposes, both turbine and generator responses are neglected to derive a simple expression for the time response.

Considering an LFC model in **Figure 11**, the model in steady-state condition at $\Delta P_L = 0$ will be a third-order transfer function as written in Eq. (47):

$$\Delta f = \left[\frac{K_g K_t K_p}{\left(\frac{K_g K_t K_p}{R} + (sT_g + 1)(sT_t + 1)(sT_p + 1) \right)} \right] \frac{-\Delta P_L}{s} \quad (47)$$

where K_g, K_t, and K_p are a constant of the governor, turbine, and power system, respectively.

Typically in a LFC system, power system time constant is relatively higher than governor and turbine time constant as described in Eq. (48). Therefore both governor and turbine time constant are negligible, and by adjusting $K_g K_t = 1$, Eq. (47) is simplified to be a first-order system in Eq. (49):

$$T_g < T_t < < T_p \quad (48)$$

$$\Delta f = \left[\frac{K_p}{\left(\frac{K_p}{R} + (sT_p + 1) \right)} \right] \frac{-\Delta P_L}{s} \quad (49)$$

The first-order system effectively reduced the unexpected response. An example of system responses of first-order and third-order systems is given in **Figure 12**.

The dynamics response of a frequency control means how the frequency is immediately corrected after the disturbance and before the system reaches new steady-state condition [17]. In this case, a static first-order model may not solve the LFC problem since the disturbance is fluctuated by the time. Therefore an adaptive model of first-order system is suggested to be used in solving this problem.

3.5 Computational procedure

Power system model is expressed in a first-order lag system. The way to build the model can be a combination of either inertia constant and damping or a gain and time constant. In this case, the second combination will be used to provide the power system model using the least square method. The power system equation can be written as follows.

Block diagram of the adaptive LFC system is shown in **Figure 13** as rewritten from [1]. The adaptive internal model is utilized to provide an update model of the power system, and the model is generated by using a LSM method. The proposed adaptive LFC controller uses MPC controller as its main controller combining with an adaptive efferent model as the internal model.

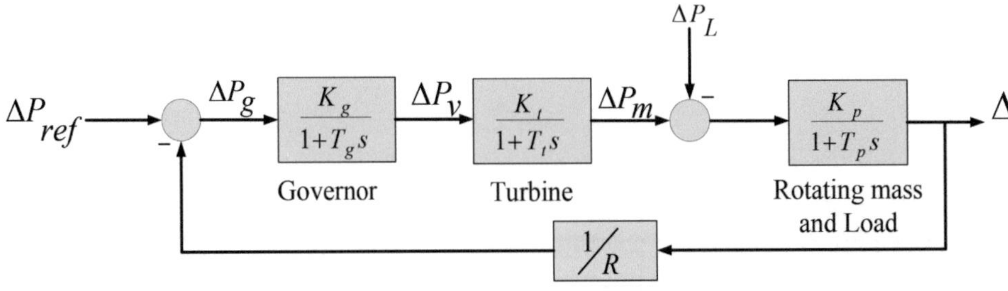

Figure 11.
Power system model.

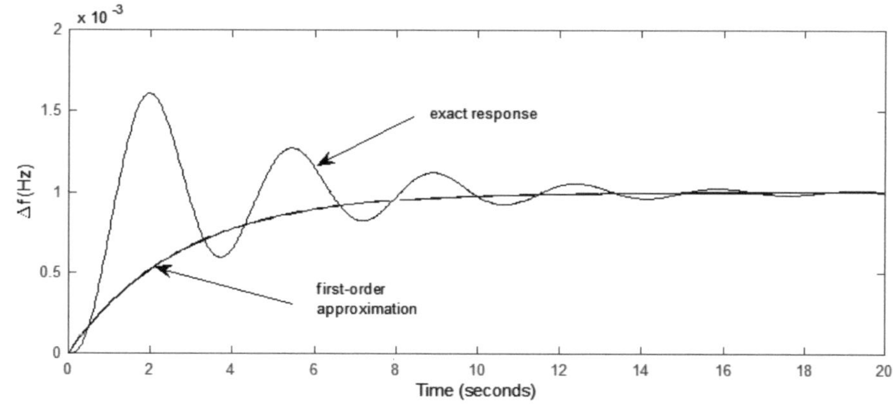

Figure 12.
First- and third-order system responses.

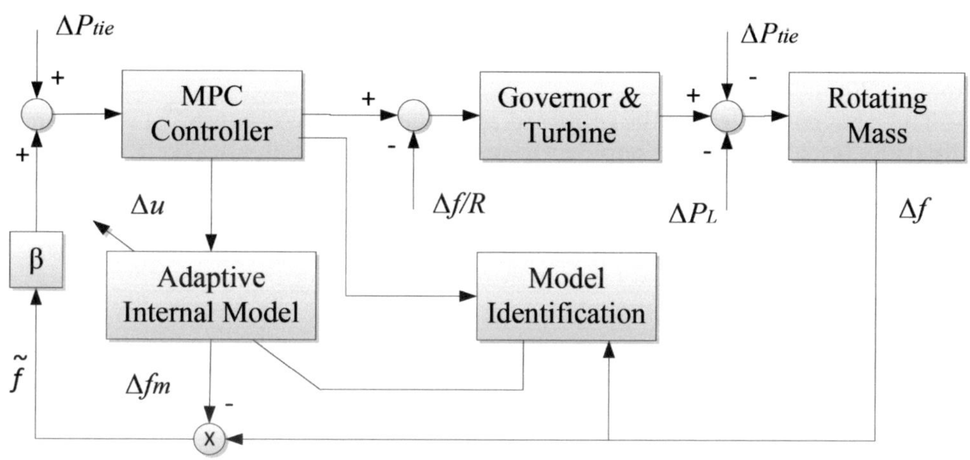

Figure 13.
Adaptive LFC system using IMC structure.

Algorithm 1:
IMC with adaptive model (type 1)

```
set disturbances and noises
configure presetting model
    ⋮
for j = 1 to simulation time
    ⋮
    for i = 1 to n-area
        calculate Δu
        upgrade state matrix Δ
        if j = desired time then
            update K & T by LSM
            change internal model
        end
        calculate model output Δym
        correct signal Δ = Δy − Δym
        ⋮
    end
end.
```

Algorithm 2:
IMC with adaptive model and controller (type 2)

```
set disturbances and noises
configure presetting model
    ⋮
for j = 1 to simulation time
    ⋮
    for i = 1 to n-area
        calculate Δu
        upgrade state matrix Δ
        if j = desired time then
            update K & T by LSM
            change internal model
            define new MPC gain K_mpc
        end
        calculate model output Δym
        correct signal Δ = Δy − Δym
        ⋮
    end
end.
```

In order to reach the best controller solution, the simulation will be done using the given algorithms. IMC type 1 includes the adaptive internal model, while in IMC type 2, controller gain K_{mpc} is also updated with the change of internal model. Therefore, IMC type 2 uses an algorithm of both adaptive internal model and adaptive MPC controller.

4. Controller test

The investigated power system consists of a two-area power system which is modified from a three-area poser system in [2–4, 9, 18]. The controller performance is tested by comparing a classical MPC controller and the proposed adaptive IMC controller. System configuration is based on **Figure 7** where the system parameter is shown in **Table 1**.

The proposed adaptive LFC controller type 1 and type 2 will be computed as follows. The gain K from P_C to Δf in the detailed model is set to 66.5, which is presetting the value for each area. An initial value of the time constant T is chosen as the same value as the power system model of **Figure 14**, which is set equal to 1. Those will be updated online based on system identification method starting from t = 20.

The main controller for the proposed controller is an MPC controller which at initial condition is set as the same as the existing controller. Prediction horizon Np = 10 and control horizon Nc = 2 are applied to both existing and proposed controllers. In the whole of simulation period, the existing MPC controller gain is not change where computed off line in the beginning before simulation is started.

Area	D [pu/Hz]	$2H$ [pu s]	R [Hz/pu]	T_g [s]	T_t [s]	β [pu/Hz]	T_{ij} [pu/Hz]
1	0.016	0.2017	2.73	0.06	0.44	0.3827	T_{12} = 0.20
2	0.015	0.1247	2.82	0.07	0.30	0.3692	T_{21} = 0.20

Table 1.
System parameters.

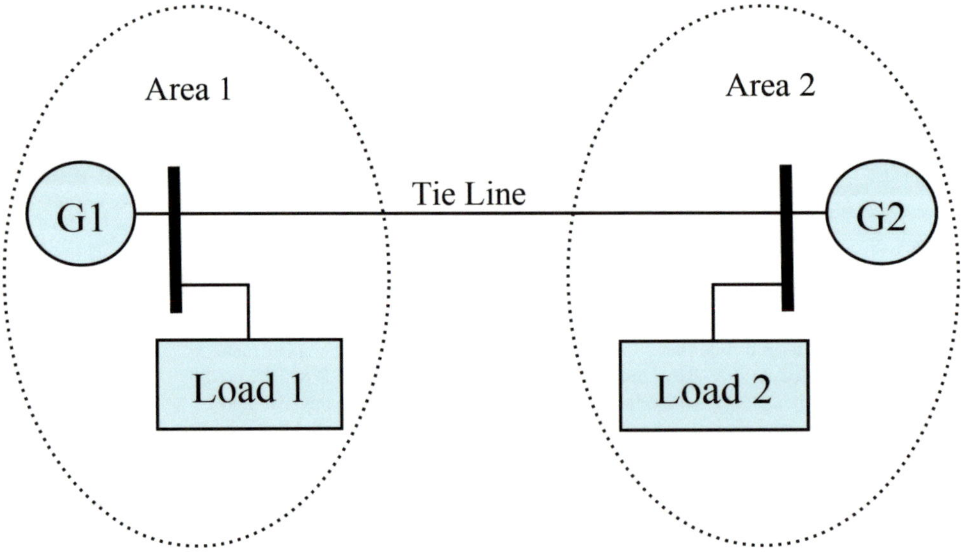

Figure 14.
Multi-area power system configuration.

Case	Disturbance	
	Random	**Step change**
1	Applied	Not applied
2	Applied	0.2 pu.
3	Applied	−0.1 pu.

Table 2.
Disturbance setting.

Three different cases are performed in simulations under disturbance setting as in **Table 2**. The disturbances are load changes which are imposed by white noises. The random disturbance is white noises with a maximum magnitude about 0.05 pu, while the step disturbance is a sudden change of load. The step disturbance is assumed as the action of economic load dispatching which is applied at t = 40 for case II and case III in addition to the white noise.

4.1 Simulations

A. Case 1.

Simulation results are shown in **Figure 15** for Δf responses and in **Figure 16** for ΔPm responses. In this case, only random disturbance is applied. It is observed that the adaptive controllers both type 1 and type 2 show slightly better performances compared to the existing MPC:

B. Case 2.

The step disturbance about 0.2 pu is applied at t = 40 s in addition to the random disturbance. The results are shown in **Figure 17** for Δf responses and in **Figure 18** for ΔPm responses.

A better performance is observed. It is noted that although the adaptive controllers are a much simpler configuration compared with the existing controller, the control performance is even better. This is an advantageous feature of the proposed method.

C. Case 3.

In this case, step disturbance about −0.1 pu. which is applied at t = 40 s assumed that load is released at t = 40 s including the random disturbance. The results are shown in **Figure 19** for Δf responses and in **Figure 20** for ΔPm responses.

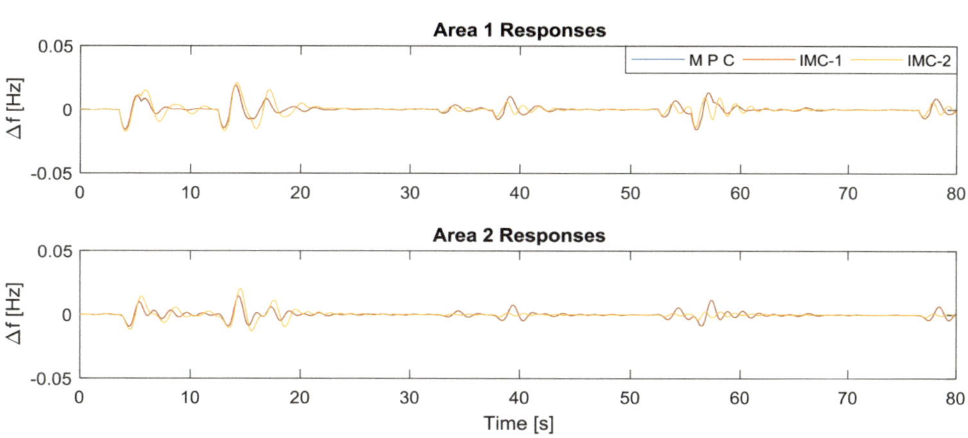

Figure 15.
Case 1: Δf responses.

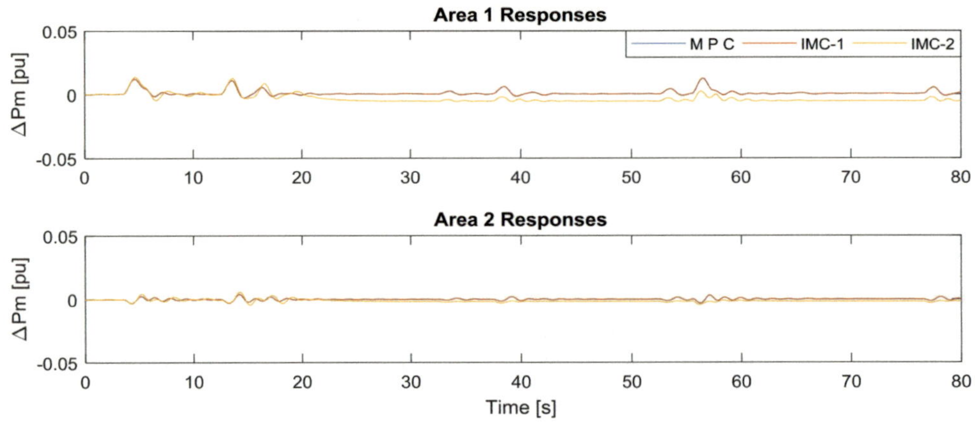

Figure 16.
Case 1: ΔP_m responses.

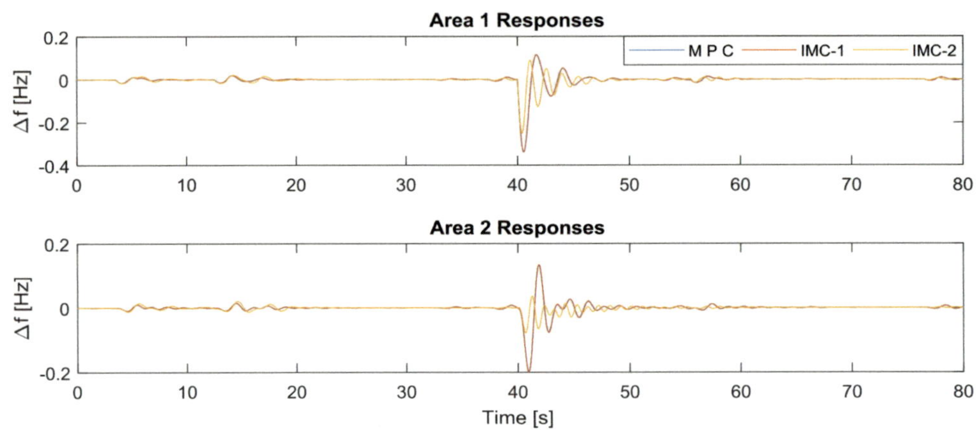

Figure 17.
Case 2: Δf responses.

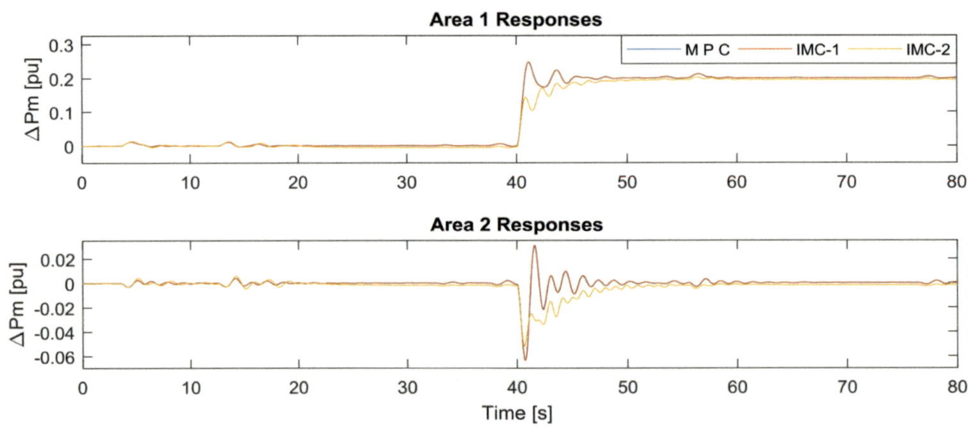

Figure 18.
Case 2: ΔP_m responses.

The similar performance is shown. It is noted that the adaptive controllers can handle incoming disturbance and also released disturbance with even better performance compared to the existing controller.

4.2 Evaluations

Figures 21–24 show how the internal model parameters are identified. It is observed from **Figures 21** and **23** that initial gain K is updated as soon as the model identification process is completed at t = 20, which are consistent values based on the LSM. In the same way, **Figures 22** and **24** show that the initial time constants T are updated very slightly around 1.0. Those values are continuously updated around the converged values.

LFC system based of IMC-1 is equivalent to the existing MPC controller in some cases, while the performance of IMC type 2 is even better in all areas and all cases. **Tables 3** and **4** list measured values of the overshoot for the step disturbance and the standard deviations of both frequency and prime mover responses for all cases. It is seen from the tables that the performance of the proposed adaptive LFC controllers is equivalent or better than the conventional MPC controller. This implies that the adaptive LFC controllers can successfully identify the target model

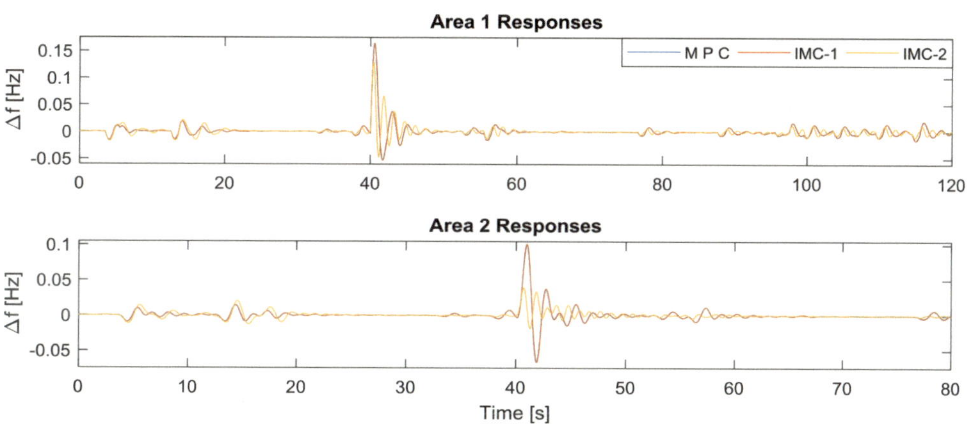

Figure 19.
Case 3: Δf responses.

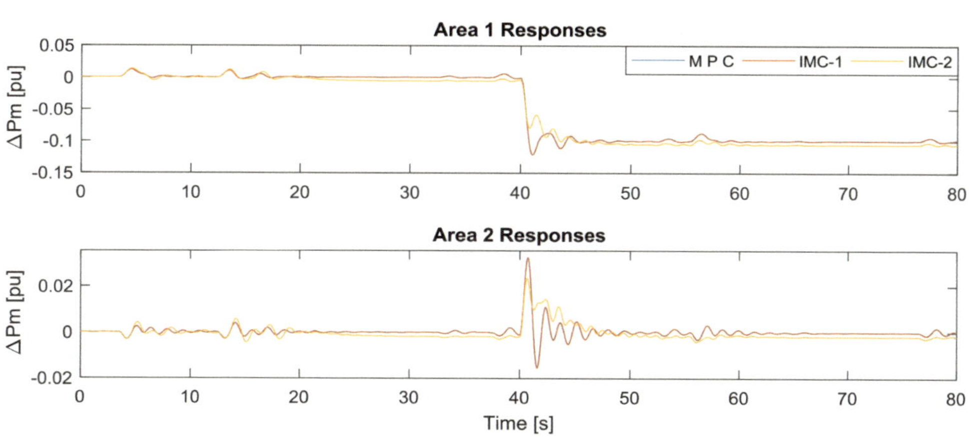

Figure 20.
Case 3: ΔP_m responses.

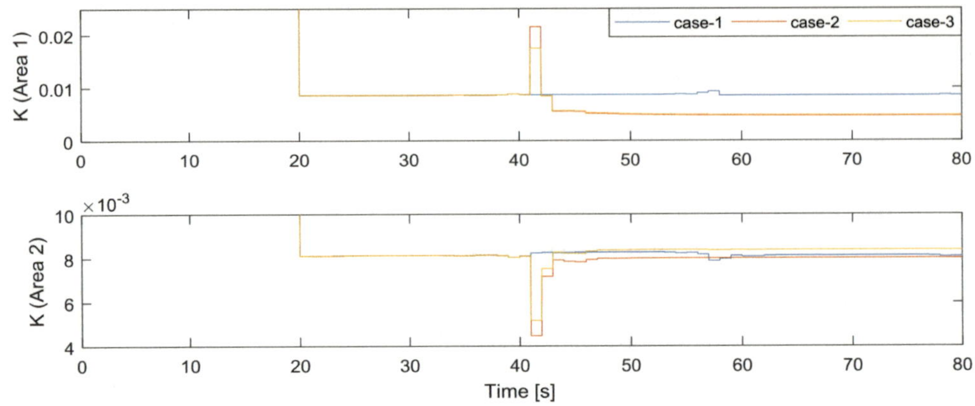

Figure 21.
Gain K *of IMC type 1.*

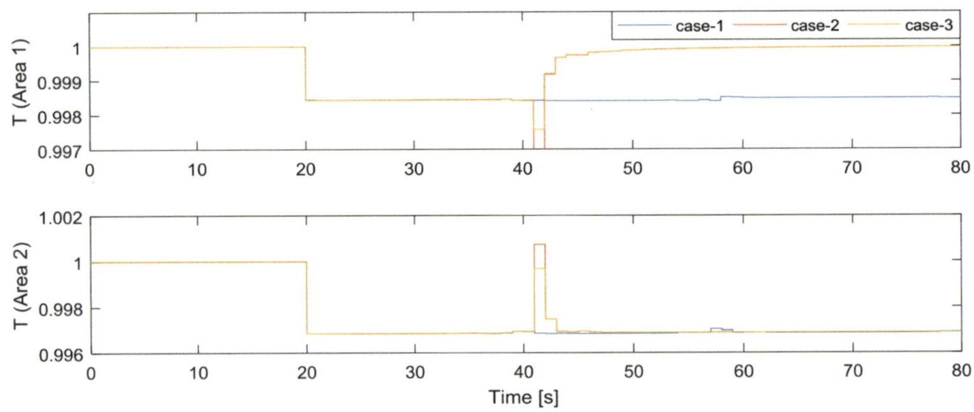

Figure 22.
Time constant T *of IMC type 1.*

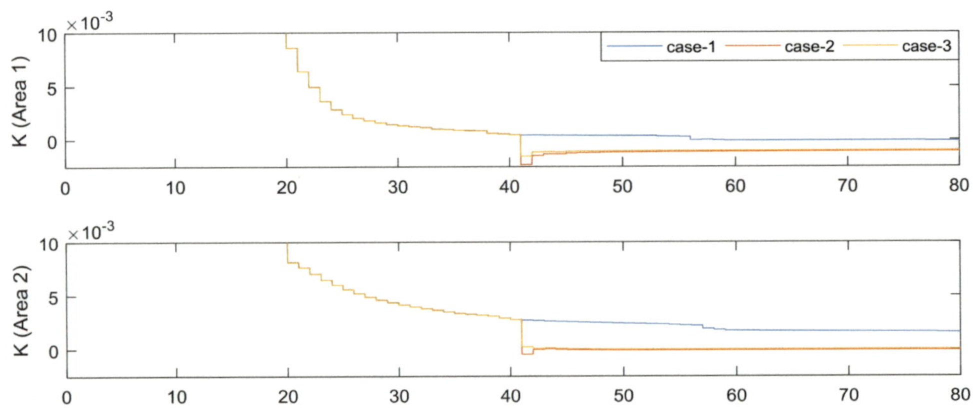

Figure 23.
Gain K *of IMC type 2.*

and handle the power system disturbances. In the same way, the controllers keep the system conditions successfully at the set points.

The simulations are carried out on PC with Intel Core i7 1.8 GHz CPU and 4 GB RAM using MATLAB 2016a under Windows 10. CPU times for the computation of

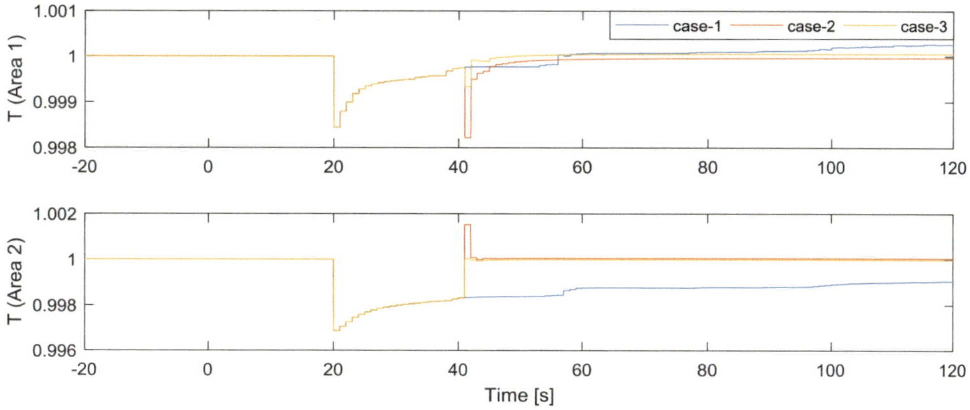

Figure 24.
Time constant T *of IMC type 2.*

Properties	Controller type	Case 1		Case 2		Case 3	
		Area 1	**Area 2**	**Area 1**	**Area 2**	**Area 1**	**Area 2**
Overshoot	MPC	0.1143	0.1339	0.1643	0.1008	0.0194	0.0143
	IMC-1	0.1139	0.1348	0.1643	0.1009	0.0193	0.0144
	IMC-2	0.0891	0.0362	0.1295	0.0393	0.0210	0.0203
Standard deviation	MPC	0.0186	0.0119	0.0094	0.0062	0.0033	0.0022
	IMC-1	0.0186	0.0120	0.0094	0.0063	0.0033	0.0022
	IMC-2	0.0131	0.0050	0.0073	0.003	0.0031	0.0019

Table 3.
Frequency responses.

Properties	Controller type	Case 1		Case 2		Case 3	
		Area 1	**Area 2**	**Area 1**	**Area 2**	**Area 1**	**Area 2**
Overshoot	MPC	0.2472	0.0302	0.0121	0.0325	0.0123	0.0039
	IMC-1	0.2472	0.0309	0.0122	0.0325	0.0124	0.0040
	IMC-2	0.2027	0.0058	0.0136	0.0236	0.0136	0.0058
Standard deviation	MPC	0.0945	0.0034	0.0466	0.0018	0.0018	0.0006
	IMC-1	0.0946	0.0034	0.0466	0.0018	0.0018	0.0006
	IMC-2	0.0913	0.0041	0.0485	0.0020	0.0028	0.0011

Table 4.
Prime mover responses.

Case	MPC	IMC-1	IMC-2
1	13.3645	16.6271	22.0835
2	13.6594	17.0485	23.0585
3	13.6208	16.7182	22.2260

Table 5.
Simulation time (s).

controllers are listed in **Table 5**. It is observed that the adaptive LFC controllers are slower than the conventional method.

In today's power system condition, where smart grid and energy management system are applied, the power system dynamics is varying. Therefore an adaptive model would be an advantage and is required for the better performance of LFC system.

It is observed that time consuming is increased together with the improvement of the controller performances. Therefore the IMC type 2 controller needs more time to compute and update both internal model and controller gain while giving a better performance. On the other hand this time consuming is the order of millisecond and so it is acceptable for an LFC system which in general operated in the order of second. Therefore the proposed controller has bit complexities in hardware for computing the model and updating the controller since its consuming time is about 40% higher than the existing MPC controller.

5. Conclusions

This chapter presents adaptive LFC methods based on IMC controller structure, where the internal model is adaptively updated online in IMC type 1, while both internal model and MPC controller gain are restructured in IMC type 2 by using the least square method. The performance of the controller is fair in handling load disturbances in spite of relatively slow control cycle with ramp rate constraints of actual systems.

Simulation results show that the gain and time constant of the internal model have been adaptively changed. This change has guaranteed the better performance of the proposed controller. Based on the system responses, the adaptive IMC controller type 1 has responses similar to the existing MPC controller, while the adaptive IMC controller type 2 has shown its superiority compared to MPC controller and IMC type 1 controller. In contrary, consuming time becomes larger by the enhancement of the controller performance.

Author details

Adelhard Beni Rehiara[1*], Naoto Yorino[2], Yutaka Sasaki[2] and Yoshifumi Zoka[2]

1 Electrical Engineering Department, University of Papua, Indonesia

2 System Cybernetics Department, University of Hiroshima, Japan

*Address all correspondence to: a.rehiara@unipa.ac.id

IntechOpen

© 2020 The Author(s). Licensee IntechOpen. Distributed under the terms of the Creative Commons Attribution - NonCommercial 4.0 License (https://creativecommons.org/licenses/by-nc/4.0/), which permits use, distribution and reproduction for non-commercial purposes, provided the original is properly cited. (cc) BY-NC

References

[1] Rehiara AB, Chongkai H, Sasaki Y, Yorino N, Zoka Y. An adaptive internal model for load frequency control using extreme learning machine. TELKOMNIKA. 2018;**16**:1-6

[2] He C, Rehiara AB, Sasaki Y, Yorino N, Zoka Y. The Application of Laguerre Functions Based Model Predictive Control on Load Frequency Control. In: the International Conference on Electrical Engineering (ICEE2018). Seoul, Republic of Korea; 2018. pp. 1190-1195

[3] Rehiara AB, Sasaki Y, Yorino N, Zoka Y. A performance evaluation of load frequency controller using discrete model predictive controller. In: 2016 International Seminar on Intelligent Technology and Its Applications. Lombok, Indonesia; 2016. pp. 659-664

[4] Rehiara AB, Chongkai H, Sasaki Y, Yorino N, Zoka Y. An adaptive IMC-MPC controller for improving LFC performance. In: 2017 IEEE Innovative Smart Grid Technologies—Asia. Auckland, New Zealand; 2018. pp. 1-6

[5] Abdillah M, Setiadi H, Rehiara AB, Mahmoud K, Farid IW, Soeprijanto A. Optimal selection of LQR parameter using AIS for LFC in a multi-area power system. Journal of Mechatronics, Electrical Power, and Vehicular Technology. 2016;7(2):93

[6] Kundur P. Power System Stability and Control. USA: McGraw Hill; 1993

[7] El-Hawary ME. Electrical Power Systems. New Jersey: Prentice-Hall; 1983

[8] Mohamed TH, Morel J, Bevrani H, Hiyama T. Model predictive based load frequency control-design concerning wind turbines. International Journal of Electrical Power & Energy Systems. 2012;**43**(1):859-867

[9] Mohamed TH, Hassan AA, Bevrani H, Hiyama T. Model predictive based load frequency control design. In: 16th International Conference on Electrical Engineering. 2010

[10] Wong C, Mears L, Ziegert J. Dead time compensation for a novel positioning system via predictive controls and virtual intermittent setpoints. In: 2009 International Manufacturing Science and Engineering Conference. Indiana, USA; 2009. pp. 1-8

[11] Wang L. Model Predictive Control System Design and Implementation Using MATLAB. London: Springer-Verlag; 2009

[12] Rehiara AB, Yorino N, Sasaki Y, Zoka Y. A novel adaptive LFC based on MPC method. IEEJ Transactions on Electrical and Electronic Engineering. 2019;**14**(8):1145-1152

[13] Shigemasa T, Yukitomo M, Kuwata R. A model-driven PID control system and its case studies. Proceedings of International Conference on Control Applications. 2002;**1**:571-576

[14] Jin Q, Feng C, Liu M. Fuzzy IMC for unstable systems with time delay. In: 2008 IEEE Pacific-Asia Workshop on Computational Intelligence and Industrial Application. Washington DC, USA; 2008. pp. 772-778

[15] Skala V. Least square method robustness of computations: What is not usually considered and taught. In: Proceedings of the 2017 Federated Conference on Computer Science and Information Systems, FedCSIS. Prague, Czech Republic; 2017. pp. 537-541

[16] More AN, Kohli PS, Kulkarni KH. Simple linear regression with least square estimation: An overview. International Journal of Computer

Science and Information Technologies. 2016;7(6):2394-2396

[17] Sivanagaraju S, Sreenivasan G. Power System Operation and Control. 1st ed. Delhi: Pearson Education; 2010

[18] He C, Rehiara AB, Sasaki Y, Yorino N, Zoka Y. Model predictive load frequency control using unscented Kalman filter. In: IEEJ Technical Meeting on Power Engineering. Jeju Island, Republic of Korea; 2018

Chapter 6

Provision of Ancillary Services by Wind Power Generators

Ayman Attya and Jose Luis Dominguez-Garcia

Abstract

The current and future power systems foresee very deep penetration of renewable power plants into the generation mix, which will make the provision of ancillary services by renewables an ultimate necessity. This would be further emphasised when green power stations replace conventional power plants that rely on fossil fuels. In this context, many control methodologies could be applied to the controllers of the green generators to enable the provision of these services, mainly frequency support and voltage regulation. Most of the available models (i.e. in power system simulators) do not include such supplementary controls to provide ancillary services. Hence, this chapter exploits key examples of these controllers that proved to be efficient and widely accepted. In addition, this chapter considers their integration into the conventional controls of green generators, where the focus is on wind energy.

Keywords: frequency stability, inertia, reactive compensation, small signal stability

1. Introduction

The European Union has announced the binding objectives of 27–32% of energy to come from renewable energy sources (RES) by 2030, with an associated CO2 emissions reduction target of 40% (relative to 1990) and at least 32.5% increase in energy efficiency following the COP24 (held in Katowice) to keep global warning well below 2°. To do so, some future European energy scenarios even foresee a very high RES penetration close to 100% by 2050. Presently, the most competitive RES technologies are wind and solar photovoltaic (PV) with hydropower pumped storage and small hydropower stations to support this drift.

Common characteristics of RES technologies (except conventional hydropower) are the variability of the primary energy source and the fact that they are typically connected to the power system through power electronic (PE) converters. Therefore, RES power plants are not synchronous and hence do not contribute naturally to system inertia. Moreover, they are not mandated to provide any type of ancillary services (AS)[1], which is still valid given the high share of conventional synchronous power plants in the present power systems. Hence, it is still not clear how the system stability could be maintained with high penetration of non-synchronous RES generation and associated reduced inertia. This is already a challenge in small networks such as that of Ireland, and it is a growing obstacle in larger

[1] Ancillary services are grid support services required by the power system (transmission or distribution system) operators to maintain integrity, stability and power quality of the power system.

IntechOpen

synchronous areas of Europe (e.g. Great Britain, Central and Southern Europe and Nordic countries). Thus, serious economic consequences may result if efficient and cost-effective solutions are not identified and implemented. However, the ambitious plans of achieving very high RES penetration into the installed generation capacities (i.e. retirement of synchronous plant) will require the strong participation of RES plants to support all aspects of power system stability and security of the electricity supply.

To incorporate AS provision from RES plants in an effective manner, a series of design and operation tools must be created to identify the optimal approach to be taken. These tools, and the incorporated benchmark models, must address crucial aspects such as, PE converter interfaces, intelligent controllers, market structures, communications, and overall power system optimised operation, including system health and assets degradation.

The coordination between a wide range of RES plants and the correlated technologies (e.g. high-voltage direct current (HVDC) corridors and energy storage systems (ESS)) must be considered through comprehensive controllers, which dispatch and regulate the contributions of these assets to maintain system stability during normal conditions and severe events. Thereupon, to model, evaluate and validate such scenarios, there is the need for developing comprehensive models of RES generation units that include supplementary controllers to enable these units to provide a wide range of AS. In this context, this chapter presents the main control concepts to provide frequency and voltage support as well as oscillation damping by wind turbines and farms according to the state of the art.

2. Modelling of frequency support

One of the key roles of transmission system operator (TSO) is to maintain the balance between power generation and load demand. However, the ideal balance (i.e. zero deviation) is unrealistic due to the dynamic nature of load, which cannot be fully controlled by the TSO. Hence, there is always an allowed margin of deviation, which reflects to the power system frequency and the associated band of acceptable frequency oscillations (typically 20–30 mHz for a 50 Hz power system) [1].

For the ages of conventional power systems, where synchronous, centralised and fully dispatched generation units dominated, frequency stability has not been a problem. In other words, each generation unit has a defined role, achieved through simple controls (e.g. governors) to maintain frequency stability. This includes fine changes in frequency due to normal load dynamics, major events which could occur due to sudden loss of generation units, or network issues as transmission lines tripping. Such events initiate large deviation between generation and demand leading to severe drops/overshoots in the system frequency. The interconnected generation units have to respond as quickly as possible to these events to curtail the magnitude of the frequency deviation (frequency nadir) and rate of change of frequency (ROCOF) and restore the frequency to the safe 'deadband'. Why? Because if this does not happen, the protection relays operating on these signals will trip, disconnecting the generating units, which would excavate the event and could lead to a total blackout.

Conventional generation responds 'naturally' to any frequency deviation due to the inertia of their rotating parts. In particular, the generation units release some of the stored kinetic energy (KE) in its rotating parts, converting it to electrical energy to tackle the power imbalance (i.e. *inertia response*). The same process occurs in the case of frequency positive deviation, but the unit stores more KE (i.e. the machine accelerates). Typically, thermal power stations have a second defence mechanism, 'speed governor', which regulates the input mechanical power (e.g. from a steam turbine) to maintain the deviation between mechanical shaft speed and the synchronous electrical speed within the deadband. This is called the *primary response*. There is a second control loop, which is effective for small deviations, and it is applied to restore the frequency to the safe deadband that is the *secondary response*. After a sudden change in load, the unit is *re-dispatched* to increase or decrease its generation set point according to the new conditions (i.e. *tertiary response*). In conclusion, the conventional generation unit applies four successive mechanisms to preserve frequency stability by diminishing any incident power imbalance: (1) natural inertia response (within 5 s from the event), (2) primary response (within 30 s from the event), (3) secondary response (within 10 min of the event) and (4) tertiary response (within 1–2 h of the event).

2.1 The widely proposed concepts

The main objective of frequency support (FS) supplementary controllers is to provide and regulate a certain of responsive additional active power during frequency excursions. FS controls usually have two operation modes: normal operation (active when frequency is within safe deadband) and support operation (active during frequency events). At normal operation, the controller has to maintain a predefined amount of power reserve that could be utilised at frequency events, i.e. support operation. However, the KE extraction concept does not apply any special control strategy on the wind turbine in normal operation as explained later. The three main concepts are illustrated in **Figure 1**.

2.1.1 KE extraction

This is the most economic concept from wind farms operators' viewpoint, as it does not violate MPT at normal operation. This method relies on extracting and converting a certain amount of the stored KE in the rotating parts of the wind turbine (WT) and converts it into electrical energy, i.e. active power to tackle frequency drops [2]. This process mimics the natural inertia response of synchronous generators, which are directly connected to the AC grid and not decoupled by the power electronic interface as the case in WTs.

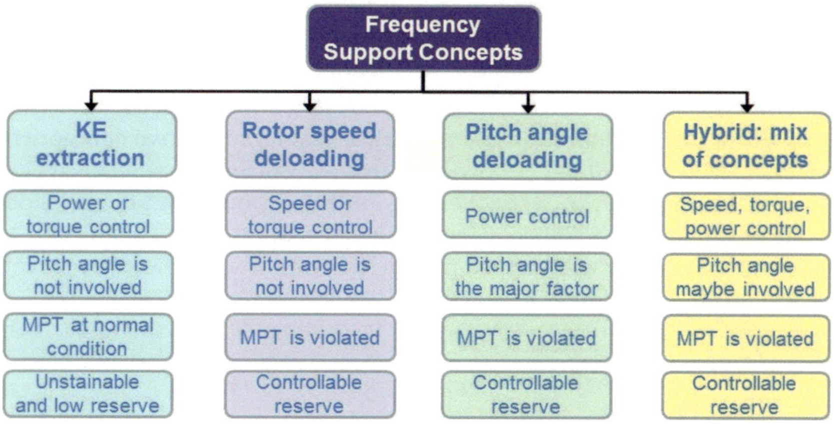

Figure 1.
The main concepts of FS in wind power generation (maximum power tracking (MPT)).

The method could be a high risk to power system stability, as the amount of extractable KE is strongly dependent on the incident wind speed, and usually this amount is rapidly depleted (2–5 s) according to the magnitude of support power and the moment of inertia of the rotating parts, mainly the rotor blades and the generator set.

The widely used control models to equip a WT with this method are focused on an inner control loop of P or PI type where a predefined constant or frequency-dependent power step is applied. Hence, the wind turbine is forced to slow down, as the input harnessed wind energy is less the electrical demand. The controller always suffers several discontinuities due to the applied limiters, e.g. on the allowed WT speed not to drop beyond a certain threshold to avoid WT complete stop. Likewise, when the WT recovers its nominal rotor speed, its output has to be regularly and slightly reduced below the available input aerodynamic energy to ensure a smooth and safe recovery to the nominal speed without major power perturbations.

2.1.2 Pitch angle deloading

This is the most applicable method used by the industry due to its simplicity, as it does not interfere with the main controls of the WT. The pitch angle (β) is the inclination of the WT blade from the axis of the incident wind speed. To harness the maximum possible wind energy, pitch angle should be zero. However, a small non-zero pitch angle would 'deload' the input wind energy to the WT. Hence, in this FS method, an amended set point is fed to the pitch angle controller to reduce the input power to the WT according to the applied deloading approach. There are two types of deloading; the first is when the input power is deloaded by a certain ratio of the available optimum input, i.e. deloading factor is a percentage, and this is called the delta deloading. The second type is to maintain a constant power reserve by reducing the input by a certain magnitude in MW, and this is called balanced deloading [3].

2.1.3 Rotor speed deloading

This method is relatively new compared to the other two concepts. It was mainly proposed to enable consistent deloading of WT output without using pitch angle control. The concept uses a P or PI controller to run the WT at a slightly higher or slower rotor speed than the reference speed produced by MPT technique. This approach has two outcomes: (1) the WT output is slightly deloaded; however, it is challenging to maintain a constant deloading ratio compared to pitch deloading. (2) The amount of extractable KE is influenced. Accordingly, it is preferable from the KE perspective to run the WT at a slightly higher speed; however this is not the favourite option from WT load and fatigue viewpoint [4].

When the WT implements overspeed deloading, at the very early interval of the frequency drop, this method provides frequency support with two components: (1) the extracted KE as the WT slows down towards the optimum rotor speed (i.e. MPT speed) and (2) the margin between the available input power and the deloaded output. However, for some control designs, this process ends up rapidly and leaves the WT without controllable reserve until the event ends, and the WT recovers the normal overspeed operation.

2.1.4 Hybridization of concepts

As expected from the title, many researchers tried to mix two or even three concepts to provide FS by wind power [5–7]. The overall objective of these trials is to avoid the drawbacks of every concept that can be summarised as follows:

- Energy wasting due to continuous deloading [8]
- Excessive mechanical loads due to continuous pitching
- Uncontrollable during the event and very short-lasting support
- Unconfirmed predefined reserve amount

The following example illustrates how the three concepts could be applied.

Example: A double-fed induction generator WT (type 3) has a rated power of 2 MW and speed control range between 0.7 and 1.2 per unit with reference to WT base rotor speed. The WT is equipped with active pitching system (β ranges from 0 to 50°). The WT applies a conventional torque-speed control to track the rotor speed that achieves the maximum power point [9].

The grid operator requires the WT to respond to frequency drops, providing an incremental positive change in its output within 0.5 s from the instant the frequency departs the safe deadband. What are the possible solutions to comply with this requirement?

Possible solutions: As a WT operator, they would need to decide the amount of support and the adopted FS concept (hint: the economic aspect is not considered in this discussion). As the grid operator requirements are so flexible, the *KE extraction* could be a reasonable option such that the WT provides 'something' when the frequency drops. In that case, the amount of reserve is not predefined but relies on the operation conditions of the WT when the frequency events occur. The simplest way to achieve this is to apply an incremental positive change in the reference torque (or power) using Eq. (1):

$$\tau_{ref} = (1 + O_F) \cdot \tau_o \tag{1}$$

where τ_{ref} is the reference torque input to the outer control loop, τ_o is the optimum torque and O_F is the overloading factor, typically 10–15%. This exceptional set point continues as long as the frequency event persists or when the WT reaches its minimum rotor speed (0.7 per unit in the given example).

Alternatively, it could be assumed that WT would provide a constant reserve of 10% of the optimum output; hence the simplest way to achieve this is to apply *pitch angle deloading*, using Eq. (2):

$$P_{ref} = (1 - D_F) \cdot P_o \tag{2}$$

where P_{ref} is the reference power input to the pitch angle controller and D_F is the deloading factor adjusted to 10% and P_o is the optimum output (all values are in per unit). The available reserve is $D_F \cdot P_o$.

Another more sophisticated solution is to maintain a constant reserve of 5% of the WT rated power, i.e. 0.1 MW. This could be achieved using *pitch angle deloading* using Eq. (3):

$$P_{ref} = P_o - \frac{D_M}{P_r} \tag{3}$$

where P_r is the rated power of the WT, namely, 2 MW and the deloading margin (D_M) is 0.1 MW. The available reserve for this approach is D_M.

Both Eqs. (1) and (2) are applied during normal operation, and when frequency violates the safe margin, P_{ref} switches to be equal to P_o; hence the pitch angle is reduced or restored to zero if the incident wind speed is below the rated wind speed of the WT.

The next subsection explains and exploits the modelling and integration of different supplementary controllers.

2.2 Modelling and integration of supplementary controls

The supplementary controllers always act on reference torque or power that are input signals to the power electronics interfacing the WT to the network (see **Figure 2** for a type 3 double-fed induction generator WT). The design of the controller relies on the adopted support method, for example, a pitch deloading controller receives the WT output power as an input and provides the incremental change in the reference power that is fed to the pitch angle controller [10] as shown in **Figure 3**. This would apply an increment change in the actual pitch angle, normally by 2–5°, to deload WT output by a certain margin.

This controller could be integrated into the DFIG library model in the Simscape Power Systems or DIgSILENT library, which is explained in the next subsection.

2.2.1 Integration of supplementary controllers to library models

The key challenge to the integration process successfully is to adapt the input and output signals of the supplementary controller to the main model. This includes the units of signals (per unit or actual), the acceptable range of each signal and the

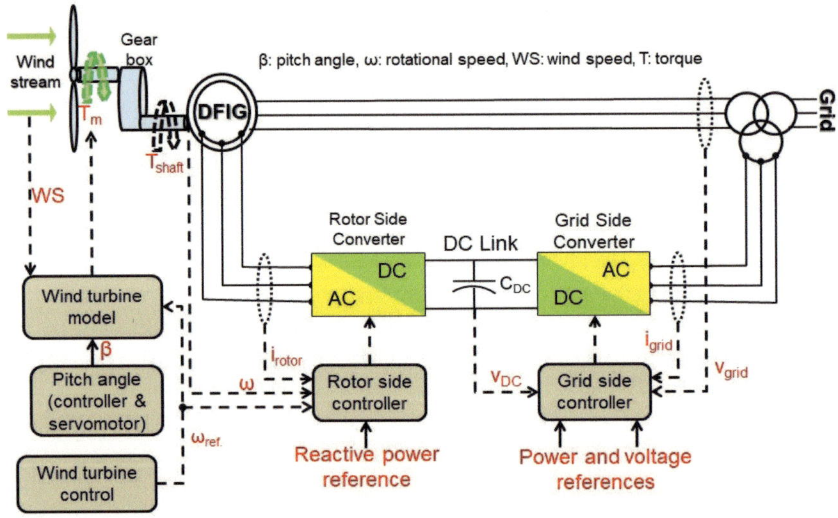

Figure 2.
Detailed block diagram of the electromechanical system of a DFIG wind turbine and the associated controls.

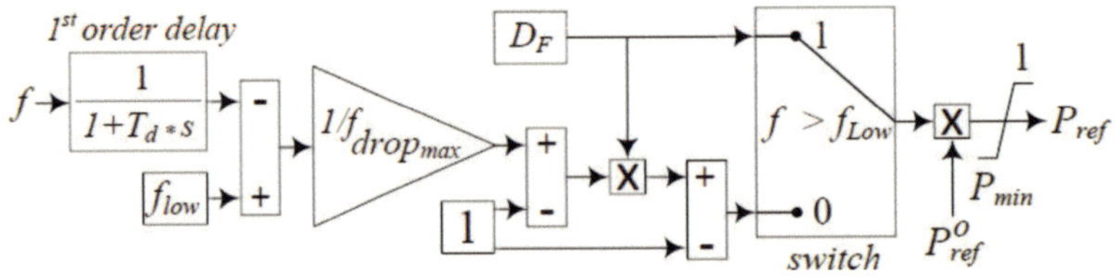

Figure 3.
Pitch deloading controller with nominal reference power as an input (D_F, deloading factor; f_{dropm}, worst frequency drop to waive the deloading; f_{low}, frequency at end of safety deadband; T_d, delay time constant).

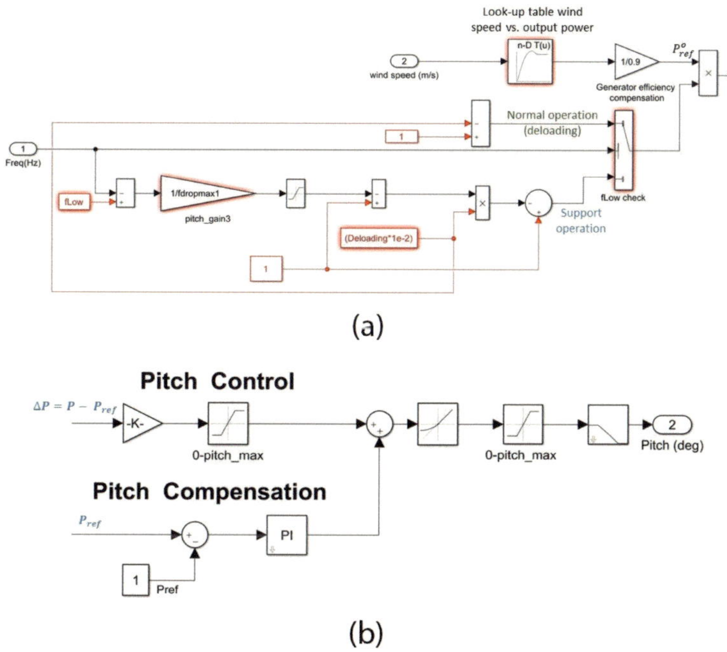

Figure 4.
(a) The Simulink model of the pitch deloading support and (b) a possible way of integration to the DFIG library model controller (the pitch angle controller).

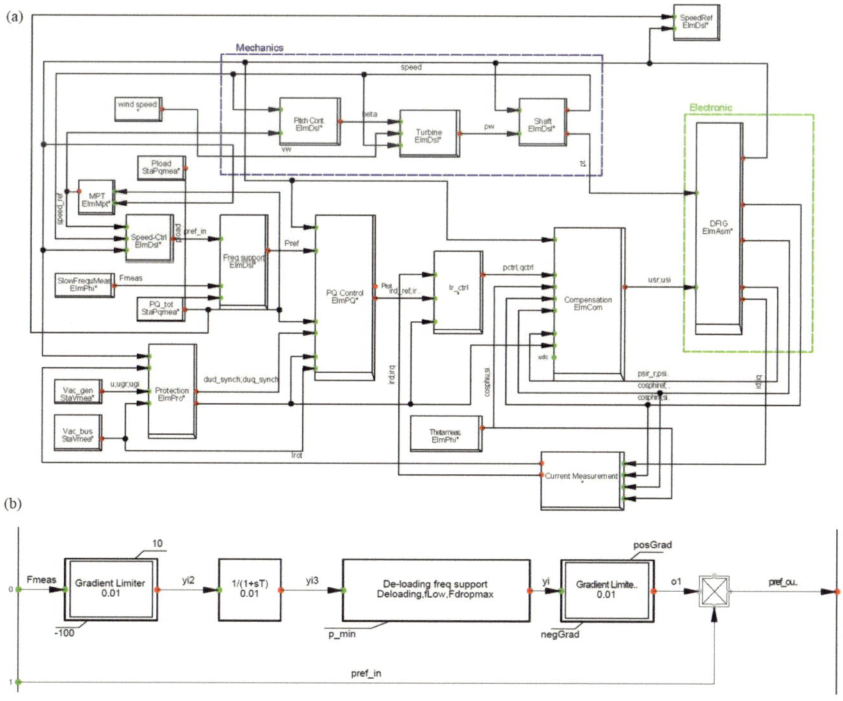

Figure 5.
(a) The Simulink model of the pitch deloading support and (b) a possible way of integration to the DFIG library model controller (the coloured frames in figure a mark the additional blocks to the generic model in DIgSILENT).

sampling time that suits the functions of the controller. For example, the controller in **Figure 3** could be integrated into the average WT model in MATLAB as illustrated in **Figure 4**.

The same control concept could be integrated into the DFIG model in DIgSILENT library as illustrated in **Figure 5**. The integration idea is simple, where the default reference power signal (P^o_{ref}) embedded within the generic frame of the WT is amended through an addition block (blue framed) using the controller illustrated in **Figure 3**. The block has two input signals, system frequency measurement at the point of common coupling of the wind farm and P^o_{ref}.

The frequency signal is obtained from an additional block (green framed), which is a standard phase-locked loop (PLL) block that can be found in the DIgSILENT library. The implementation of this support method in DIgSILENT is tested through several case studies in [11], where the response of both the WT and the connected power system is captured and analysed.

One of the challenging tasks is to tune the parameters of the PLL [12] to achieve an accurate and clean frequency measurement, so that it does not affect the performance of the controller negatively. Hence, there is a strong research trend towards support methods that do not require frequency measurement [13].

3. Modelling of voltage support

The provision of reactive power/current support is not as challenging as active power support (frequency issues and generation-demand balance), as it does not require securing power reserves. However, it is a very critical task due to its execution within very short time (milliseconds) compared to frequency support, mainly during faults. In addition, grid code requirements are always very strict in this regard; hence it could be challenging for WTs to comply. An interesting question may arise: Why are voltage requirements more critical and restrictive while one of the most key issues of power system stability is the active power balancing? The answer is already implicit in the question, which is because of the power balancing, as these requirements ensure that generation units stay connected to the grid. Hence, these units would continue generating power to the grid as soon as the fault is cleared, avoiding any consequential power imbalance.

3.1 Grid code requirements of voltage regulation

Regarding voltage support, the main objective of a grid code is to define when the generation unit is allowed to disconnect, commonly known as fault ride through requirement. As an illustration, and as shown in **Figure 6**, the generator must be kept connected as long as the minimum voltage (V_{fault}) is sustained for a duration shorter than t_{clear}, which is the clearance time of the fault. Likewise, the relays sensing the rate of change of voltage must be tuned to accommodate the post-fault rate of voltage recovery (from t_{clear} to t_{rec1}). The recovery could halt where a low-level voltage sustains until t_{rec2}; however, the generator must be kept connected within the defined time span and so on. This pattern differs from one system operator to another; in some cases, the intermediate recovery phase is not included to allow higher tolerance [14, 15]. The typical values of the pivot voltage and time points of this pattern are summarised in **Table 1**. This should be the first part of compliance, where the second part is related to the provided support to voltage recovery to the acceptable margin (i.e. typically 1 ± 0.1 per unit). According to the majority of grid codes [16, 17], the generation unit should maintain a 1 per unit reactive power/current injection during voltage dips, and it reduces gradually in relation to the voltage

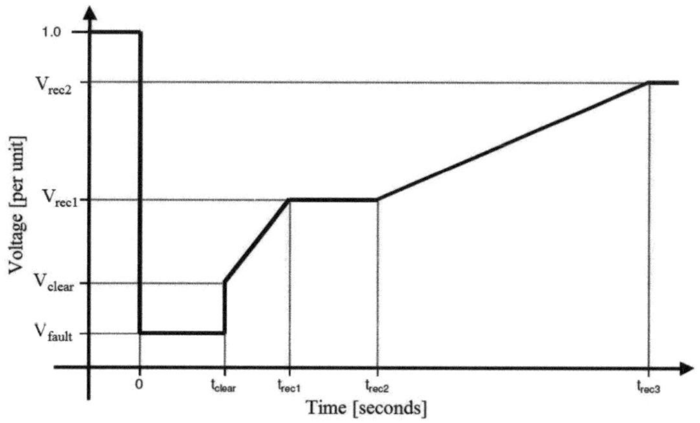

Figure 6.
Generic low voltage ride through (LVRT) grid code requirements.

Voltage limits	Value	Time limits	Value
V_{fault}	5–30%	t_{clear}	0.14–0.25 s
V_{clear}	70–90%	t_{rec1}	$t_{rec1} \geq t_{clear}$
V_{rec1}	$V_{clear} < V_{rec1} < V_{rec2}$	t_{rec2}	$t_{rec1} < t_{rec2} < 0.7$ s
V_{rec2}	85–95%	t_{rec3}	$t_{rec2} < t_{rec3} < 1.5$ s

Table 1.
Reference parameters during frequency events.

recovery. Some grid codes define the required pattern of the injected reactive current at different voltage levels, similar to the main ride through curve; however, it is more accurate to define the reactive current rather than the reactive power as the voltage dip mitigates the capability of reactive power transmission; hence the current value is more reflective and critical.

3.2 Modelling and integration of supplementary controls

There are three main solutions that enable the WT to ride safely through voltage dips; these solutions require the connection of additional equipment to the WT, and it differs based on the WT type; however type 3 is brought to focus in this chapter. The first solution is the dominant one, namely, a *crowbar circuit*, connected between the rotor-side converter (RSC) and the rotor windings of the induction machine of a DFIG. According to the applied technology of the converter, either IGBTs or an advanced voltage source converter (VSC), in addition to the ratio between stator and rotor voltages, the presence of a step up transformer between the RSC and rotor windings is decided. However, the modern designs avoid the presence of this transformer to mitigate the size and cost of the WT. The crowbar has different topologies and a three-phase resistive load to dissipate the additional energy during the fault and provides an alternative path for fault currents bypassing the RSC. The same concept can be applied using dc resistive load connected via a three-phase bridge [18]; however the crowbar circuit is one of the drawbacks of the DFIG compared to the PMSG full-rated converter type 4 WT [19]. As an illustration, the WT losses controllability during this stage because the RSC is decoupled and replaced by the crowbar circuit to protect the WT back-to-back converter from high currents and voltages, including the dc link voltage

[20]. Hence, it is aimed to reduce the connection time of the crowbar circuit without compromising the safety of the WT. In addition, this allows to provide reactive current support earlier when the controllability of the RSC is retained. The second LVRT method is applicable for both types 3 and 4, where a dc chopper is connected across the dc link between the RSC and grid-side converter (GSC), as shown in **Figure** 7, to dissipate the additional energy and stop the evolution of the magnetic flux of the machine. However, this method is more expensive than conventional crowbar circuit [21, 22]. The third method is relatively novel, where a superconducting fault current limiter (SFCL) is connected between the RSC and GSC as shown in **Figure 8**. The SFCL operation is based on the physical nature of the integrated superconductor where it changes its conducting state from normal to superconductivity according to the material characteristics, as well as the ambient temperature and the expected current continuity to which the device is designed. There are novel topologies of SFCL which are exploited to anticipate dc faults for large-rated dc connections within very short time and with reduced current surges [23]. This chapter will consider the resistive SFCL type, which is already applied to a wide range of electrical equipment; however, it is still an immature technology in LVRT hardware of wind turbines [24].

3.2.1 Key control features

The key features of voltage support for wind, applicable to any power electronics interfaced to the grid by power electronics, are as follows:

- The triggering time: duration of sustainable fault conditions to trigger the support operation mode (typically two to three cycles).

- Connection/activation time: the time for which the LVRT equipment and/or operation mechanism remains active from the instant of triggering. It does not have a typical value, but it has two main approaches: first, setting a constant duration regardless of the fault conditions and second, observing the fault and stopping the LVRT operation after a certain period of fault clearance assurance.

- The way to sense the fault occurrence: this could be achieved by observing the voltage level at the connection point (of the wind farm), machine rotor current (in the case of a DFIG wind turbine) and the dc link voltage (the link between the GSC and RSC).

These three main features are illustrated in **Figure 9**. These features were tested through comprehensive scenarios, and their dynamic performances were critically analysed in [25]; however, this chapter is focused on the modelling aspect rather than the impact of these controls on the power system and WT.

The SFCL has not been practically deployed as a LVRT hardware in the wind power industry. However, it has a promising potential, mainly that it showed merit when it is applied in the protection of distribution networks [23].

The crowbar circuit can be modelled in different ways, where the simplest approach is to use ideal switches whose on/off signals are generated by the applied LVRT control as illustrated in **Figure 9** (The output signal is used to trigger the LVRT hardware). The crowbar circuit can have different topologies: delta-connected equal resistors or Wye-connected equal resistors or dc resistors [18]. The most challenging aspect would be the selection of the correct value of the resistor that achieves a compromise between suppressing the fault current below safety

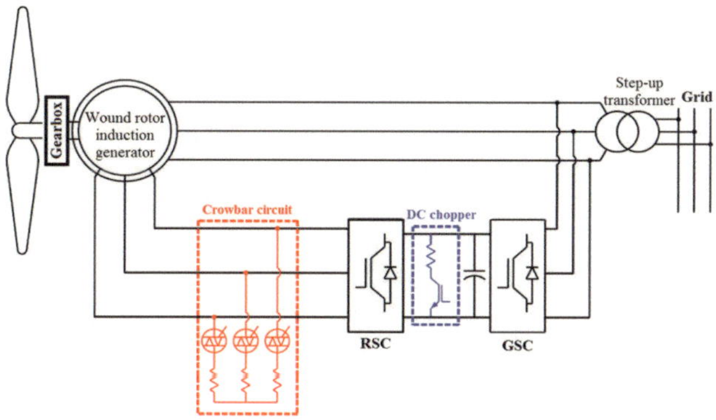

Figure 7.
Three different LVRT solutions of the DFIG type 3 (GSC, grid-side converter; RSC, rotor-side converter).

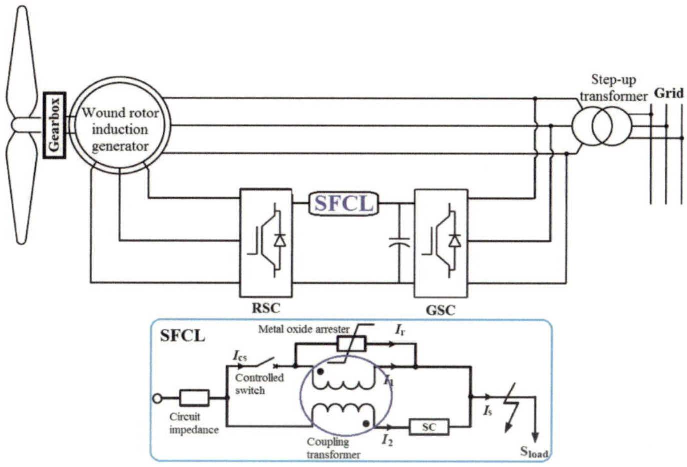

Figure 8.
Schematic representation of the reaction SFCL connection to DFIG.

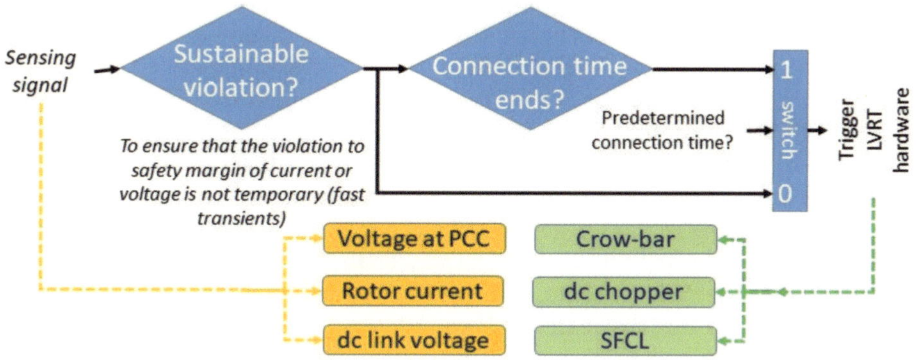

Figure 9.
The main features and their common solutions in LVRT for renewable energy units/farms (SFCL, superconductive fault current limiter).

limit, without causing excessive heat. The crowbar can be connected in series with the RSC as illustrated in **Figure 10**.

The second LVRT hardware is the dc chopper that can be simply modelled as a resistor in series with an ideal switch and connected across the capacitor of the dc

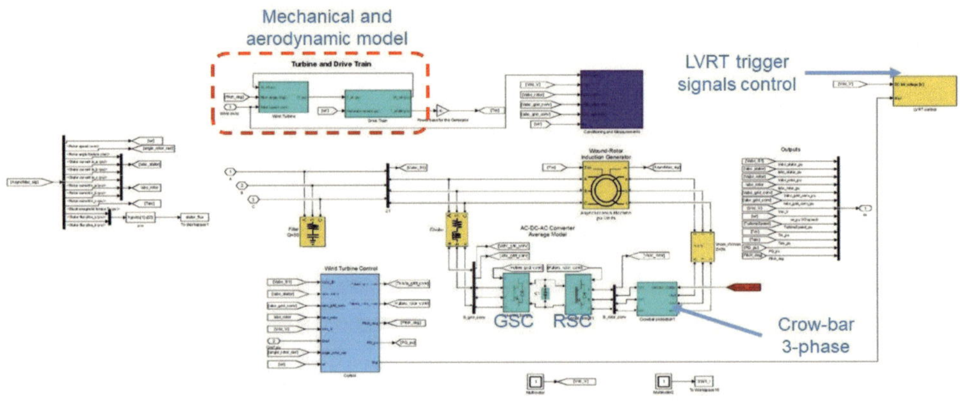

Figure 10.
An overall view of the DFIG-detailed benchmark in Simulink.

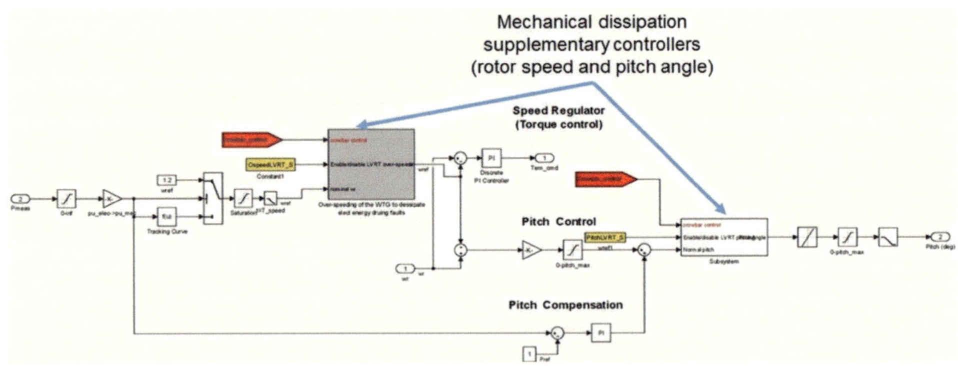

Figure 11.
The implemented supplementary controllers in DFIG-detailed benchmark in Simulink to enable mechanical aid to LVRT in chapter 1.

link between the GSC and MSC. The triggering signal to the ideal switch is provided by the applied controller, similar to the crowbar circuit. All these components are easy to find and assemble in MATLAB Simscape, where the main challenge is to set the values of the controller parameters as well as the dc chopper resistor.

A mechanical ride through method could be used, which relies only on a supplementary controller and does not require special LVRT hardware. As an illustration, the key role of protection devices is to dissipate the high fault currents through the device impedance; thus it would be helpful to mitigate the input mechanical power to the WT, in turn, reducing the generated electrical power feeding fault currents. Nevertheless, the speed of response of such mechanical methods might not be fast enough to tackle the fault currents, which will be examined through this research work. The main idea is to dissipate the input KE (i.e. wind energy) to the WT, by increasing the pitch angle to its maximum and using the excess energy to accelerate the rotor instead of causing magnetic flux evolution. However, the speed should not violate the maximum allowed limit. The pitch control retains normal conditions, and the generator speed is decelerated, such that the WT is able to resume normal power production promptly after the event [26]. An example for integrating this concept to the benchmark DFIG model in Simscape power library is depicted in **Figure 11**, where it is has two main components: the first part is responsible for slightly overspeeding the WT during the fault by implementing an increment change to the default reference rotor speed signal. The second part slightly increases the pitch angle if necessary to reduce the harnessed wind energy during a fault, hence reducing the generated electric current by the induction machine.

This method could be supplementary not the sole LVRT method, as it is not sufficient to replace the electrical solution (i.e. crowbar, dc chopper, etc.), due to its slower response; hence it could not ensure a very rapid suppression of over-currents and voltages across the WT converters, which makes the WT subject to possible risks of damage and tripping protection.

4. Modelling of oscillation damping controllers

Similar to the case of voltage support provision, oscillation damping does not require securing power reserves but proper power management and flexibility.

In the previous ancillary services presented, control variables were clearly identified to each end (i.e. frequency-active power, voltage-reactive power), while it is not the case of power oscillations. These oscillations are a natural response of the power system and/or other connected systems to any perturbation which could excite it. The oscillations can be observed in any electrical variable, including power, voltage and phase angle, among others.

Historically, the stability problem (commonly known as small signal stability) has been mainly the synchronous generators, as they are the large dominating machines governing the dynamics of power system under low renewable penetration. However, lately new causes of oscillations are evolving due to control interactions or sub-synchronous oscillations. Generally, oscillation modes can be classified depending on the systems that provoke it and their frequency, as follows:

- Inter-area: the oscillations due to an interaction of a group of generators nearby (area) with another group of generators in another area which are interconnected (typical frequency range of 0.1–0.7 Hz).

- Intra-area: the oscillations of a single generator to the rest of the system or area (typical frequency range of 0.7–2 Hz).

- Torsional: this is interactions among the mechanical electrical parts of a generator (typical frequency range: above 2 Hz).

- Control: this is related to the interplay between the controllers themselves and power system dynamics (typical frequency range: above 2 Hz).

The last two types are commonly known as sub-synchronous resonances.

Due to the importance of this stability problem, some grid codes are already requesting this service to provide oscillation damping as well as ensure that integrated controllers provide other services do not cause unexpected oscillations [27].

A common simple model for designing power oscillation enhancement controllers in conventional generation (known as power system stabilizers) is the well-known Heffron-Phillips, which represents a conventional generator connected to an infinite bus [28]; this model is built in MATLAB/Simulink as shown in **Figure 12**.

The beauty of such model is that it presents a very simple case presenting the minimum dynamics of a conventional generator allowing the design of the PSS to damp out the local (inner) oscillations. As it can be seen normally, the PSS uses the frequency or generator speed as an input and modifies the exciter voltage. The conventional control structure includes a washout filter, to ensure acting only on the desired frequency range, and after that a phase compensator (or lead/

lag block) which ensures the proper modification of the dynamic response of the generator.

4.1 Wind power to mitigate oscillation resonances

As previously stated, such oscillations could be damped or mitigated by the proper regulation of WT controllable variables. To apply this without major modifications of WT control, a supplementary controller could be integrated. From a wind power perspective, the input signal for the supplementary controller could be any signal that ensures observability of the oscillations (e.g. active power, frequency, phase angle of synchronous generators, voltage magnitude and/or phase) and produces a signal which could impact power flow within the power system (e.g. active and/or reactive power reference of WT and voltage at the connection point, among others).

It is worth to indicate that the typical structure for oscillations damping in synchronous generators is known as power system stabilizers, which are designed to damp out the generator oscillations of certain frequencies by using a bandpass filter and a lead/lag control which ensures stability of the generator by taking the advantage of the phase margins (as shown in **Figure 13**). Although this methodology could be used in WTs, the control can be simplified due to the fact that wind farm does not have a direct impact on the phase margin of the generator but only on the general power flow [29].

Generally, the WT support on oscillation damping could be classified according to the type of power being regulated, i.e. active or reactive. In **Figure 14**, potential methods to regulate relevant variables are presented, using appropriate controllers to achieve small angle stability.

The damping capability of each variable is completely different. As expected, the regulation using active power has a larger impact on enhancing stability than using reactive power as a main signal [30].

4.2 Modelling and integration of supplementary controls

It is important to remark that small signal stability is usually evaluated by two main methods: linear algebra techniques (mainly, eigenvalues and eigenvectors) and time-domain simulations.

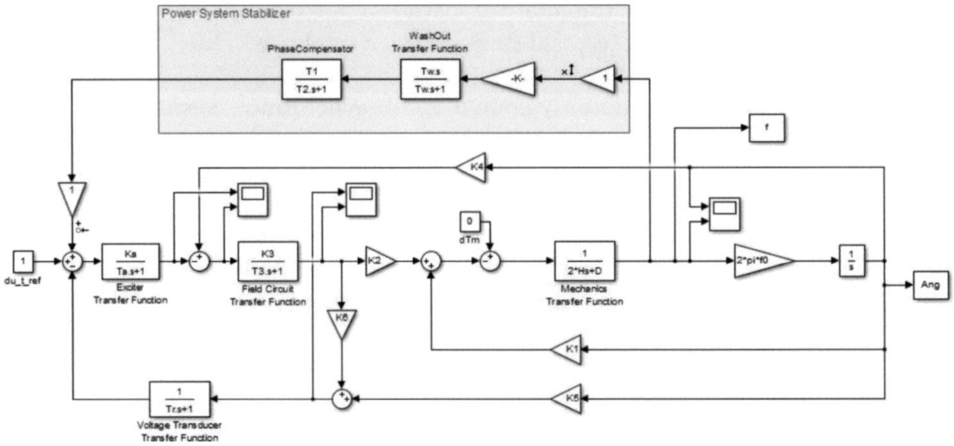

Figure 12.
Heffron-Phillips model for a synchronous generator-infinite bus model.

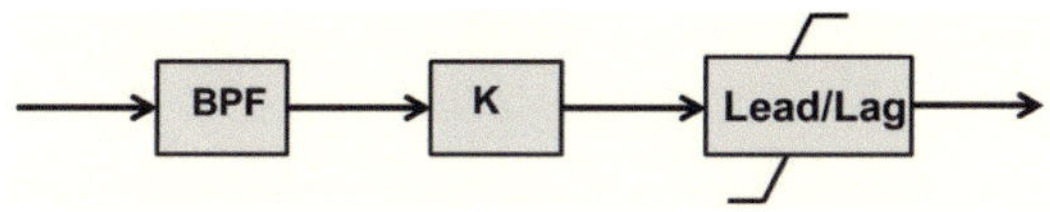

Figure 13.
Conventional PSS scheme (BPF, bandpass filter).

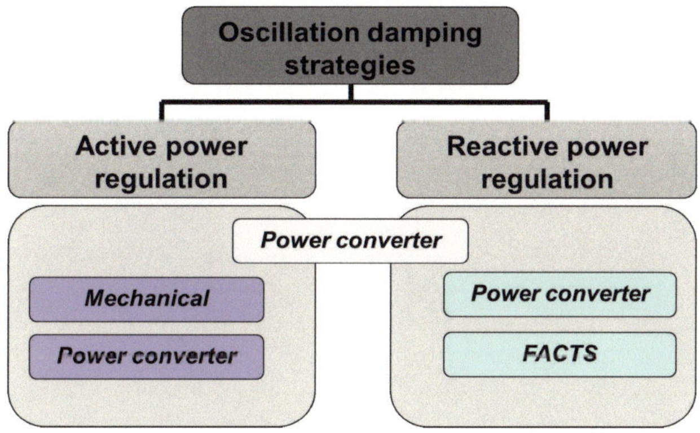

Figure 14.
Simple classification for oscillation damping provision.

The first method, based on a linearized system mathematical model, provides a simple method to identify the oscillatory dynamics within the system through the eigenvalues (i.e. $\lambda = \sigma \pm j{\cdot}\omega$). In this case, ω refers to the frequency of the oscillation ($\omega = 2{\cdot}\pi{\cdot} f$); and σ refers to the non-oscillatory part, which indicates the stability (negative = stable and positive = unstable), as well as the damping ratio ($damping = -\sigma/(\sigma\hat{}2 + {\cdot}\omega\hat{}2)$).

One common model benchmark for the power oscillation analysis is known as the Kundur (two-area model) [31].

In order to analyse the impact and contribution of wind power, the supplementary controllers could be integrated into comprehensive model which includes a detailed wind farm (each wind turbine represented by a separated mode) or aggregated (wind farm represented by one wind turbine of equivalent rating); an example is shown in **Figure 15**. This model is developed in MATLAB/Simulink environment which already included different built-in models. It is worth noting that the wind farm is based on the GE 3.6 MW model [32], which include active and reactive power control and voltage regulator, among other control systems, as shown in **Figure 16**.

The supplementary controls have been applied to the main control loops to adjust the reference values accordingly. The impact of the supplementary controller is clear at the GSC of the WT, as it is responsible for the interaction with the power system. It is of note that in the case of DFIG WT, the maximum reactive power limited by the generator rating not the grid-side converter rating.

Finally, from the modelling and control design perspective, one advantage of using MATLAB/Simulink for these studies is the integrated toolboxes for linearization of the whole system, which include system state identification and simplify the control design and oscillation detection. With the system identification methods (from the linear perspective), the user could select the desired input and output of the plant and identify the transfer function that links these signals. Linear transfer functions help to produce the state-space model; hence the eigenvalues and eigenvectors can be easily computed to obtain the corresponding oscillation modes and

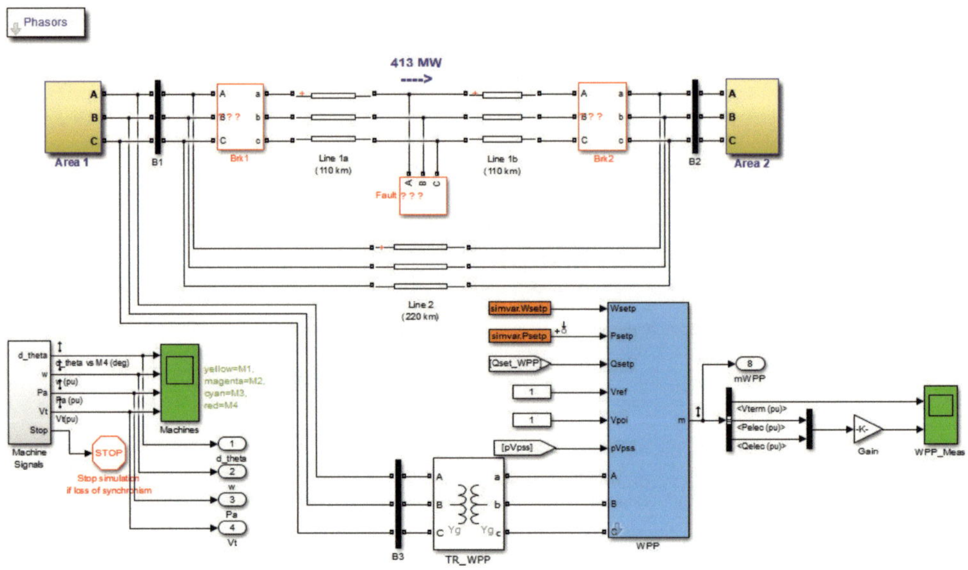

Figure 15.
Adaptation of Kundur's model by including wind power plants in MATLAB.

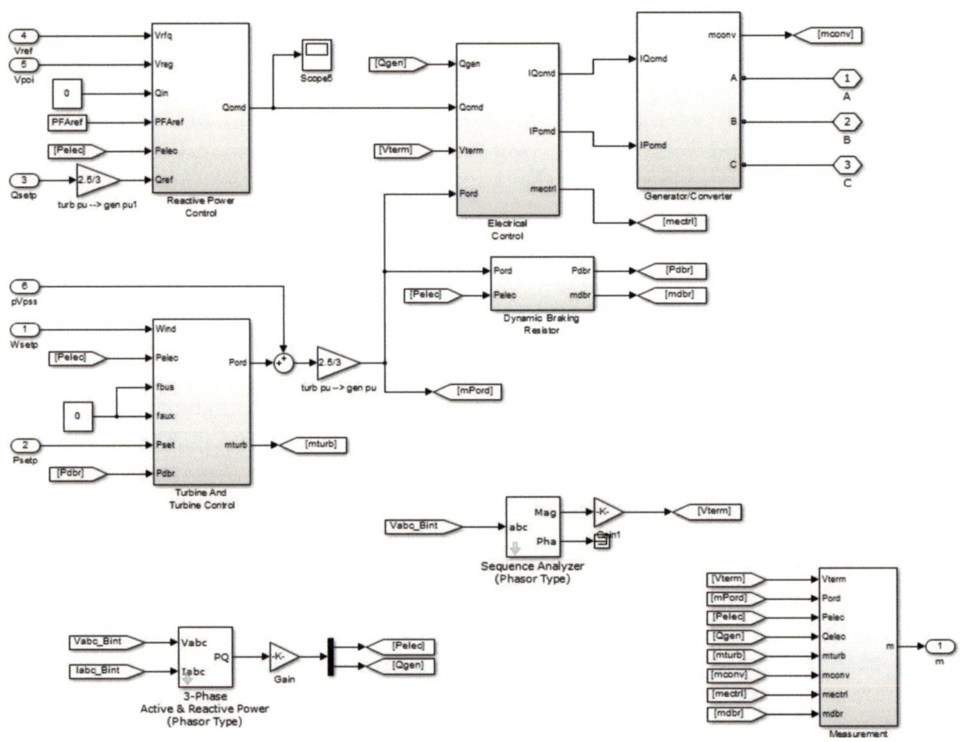

Figure 16.
Wind power plant based on GE-3.6 simulation model.

their main causes. Finally, with the state-space model or the transfer function, any type of control which allows modifying the desired dynamics could be applied, for example, pole placement, root locus and other advanced techniques as H-infinite. In case of implementing this type of studies to different simulation software as DIgSILENT PowerFactory (previously presented), it has a toolbox for eigenvalues and oscillation mode identification (computed internally by the software), but in the case of model identification and control design, it could be linked with MATLAB and python through some existing scripts.

5. Discussion

This section describes the key simplifications and potential limitations to apply these methods in the real world. It also provides a brief discussion for the main applicability barriers and practical limitations for the represented control methods.

5.1 Reasonable simplifications and potential limitations

It is worth noting that all simplifications introduced before are commonly accepted and applied in order to simplify control design and development in an acceptable model which does not include additional complexity which could slow down simulations without providing significant additional information.

5.1.1 Wind turbine/farm wise

The proposed modelling solutions are subject to some simplifications where the model parameters of the wind turbine are considered constant; however this is not ideally accurate as changes in some parameters, due to operation conditions, could lead to considerable drift in the performance of the WT when it provides a certain service. For example, the WT inertia is always seen as a constant value (1.5–3 s) according to its size and gearbox technology, but actually this inertia suffers marginal changes subject to the incident WS and the mechanical characteristics of its blades. A change in the inertia would impact the amount and duration of the provided support power during frequency drops.

In addition, one of the widely-used simplifications is to ignore the modelling of the power electronic interface (i.e. rotor-side and grid-side converters for both types 3 and 4 WTs). In fact, many power system researchers consider the power electronic interface as an ideal box with zero-time delay, where the required set-points (amended reference power, torque or speed), which are produced by the ancillary services controllers, are well received and applied by this interface. In real world, this could have minor implications; however, these interfaces are very efficient (98–99.5%), and the induced delays do not exceed a few milliseconds. This assumption is perfectly acceptable for frequency stability studies, as frequency dynamics occur within a much larger time scale (the most relaxed ROCOF restriction is 0.5 Hz/s).

Pitch angle actuator could be also a challenging aspect for modelling the WT response for pitch deloading techniques. Most of the literature considers only the delay of the servomotors, ignoring the elapsed time to move the blades (i.e. inertia of the blades as rotating masses). However, this is very minor as the blades are not moved from stationary status. Additionally, it is not easy to obtain the accurate/authentic parameters of the empirical equation which describes the variations of harnessed power against tip-speed ratio and pitch angle [33]. All WTs manufacturers do not provide open access to such critical information as it could reveal their unique aerodynamics designs of their blades.

Conversely, the previous three simplifications should not have any influence on voltage support except the accurate modelling of the power electronic interface. Actually, this interface is completely responsible for the reactive power compensation, and hence it should be modelled as accurate as possible. However, some researchers simplify the PQ limitations of the converter and set it as a square of 1 per unit for each side, which leads to 'optimistic simulations' compared to the real world.

All the previous simplifications are influential regarding the provision of oscillation damping, as this service is a complex mix between active and reactive power balancing and compensation. The oscillation modes are also sensitive to rotor inertia and dynamics as well as the capabilities of the integrated power electronic interface.

The limitations are mainly related to the expected WT response and the provided support using these models. The amount of power support (i.e. ΔP) relies to some extent on the incident WS, which is always fluctuating in contradiction to most of the models that assume that WS is constant during the event. This assumption could have a clearer influence when the amount of reserve (i.e. sustainable ΔP) is evaluated. As an illustration, in balanced deloading for example, it would be very challenging to maintain a fully constant ΔP for long durations due to the interactions of WT inertia, incident WS and different WT controls. However, in simulations this is achievable. Likewise, WS measurement is essential for many of the proposed controllers; however, in reality, this could be subject to errors and failures, where the state-of the-art technology relies on laser and could experience 0.25–0.5 m/s error [34], which should not be significant to support operation; meanwhile most of the models assume ideal WS measurement.

The assessment of the economic value of providing these services, mainly frequency support, is also limited by the accuracy of the implemented MPPT power curve which is usually provided in the vendor manual [35].

5.1.2 Power system wise

The power system main simplification and limitation at the same time are the accurate measurement and communication of system frequency to the relevant supplementary controller in the WT and/or the WF. The frequency measurement is always obtained using PLLs, and it is prone to noise and errors [36]. However, most of the implemented models in the literature applies a clean frequency signal to focus only on the merits of the proposed support methods.

The second limitation is that most of the models ignore the modelling of either the protection relays or at least their impacts. For example, the influence of ROCOF relays could be significant (stop the simulation and in reality trigger the WT protection so it comes to a complete stop) if the ROCOF threshold is violated. Many studies overcome this simplification by showing the ROCOF behaviour during the event to ensure that its presence is within the safe limits.

The same applies to voltage support, where the WT of WF converter station overcurrent relays could stop the simulation, if the overall current exceeds the limits (typically 1.4 per unit sustained for 1–3 s). This is likely to occur during symmetric faults or when the WT is operating in LVRT and suddenly switch to reactive compensation mode. In particular, as soon as the fault is cleared, the WT is required to recover the full pre-fault active power as well as maintain high reactive current to recover the nominal voltage level [13]. However, commercial simulators, e.g. DIgSILENT and PSS®E, include these protection gears or at least mimic their influence, in most of their library models.

A third key simplification is the 'ideal consistency' where all the integrated WFs models, usually a single WT of an aggregated capacity to represent each WF, are consistent in all aspects except only one or two according to the applied case study.

5.1.3 Synchronous machine infinite bus simplifications

The model presented (Heffron-Phillips) is a very simple model which mainly represents the mechanical behaviour of power system that represents the basis of frequency dynamics of the electrical network. Such simplified model neglects all electrical parts of the power systems and existing interactions among different variables as voltage, current, cable limits, etc.. In addition, as it occurs with the wind turbine/farm modelling, the delays impacting on communications and measurements must be considered when implementing such concepts for real experimentation and replicability. However, these models are widely accepted for control development.

5.2 Implementation challenges

The implementation of the proposed methods on a wide scale and in large wind power plants will face two main obstacles: data access and communication as well as standardisation. The required volume of data is massive, including models, control parameters, live measurements and signals across the coordinated assets. In addition, communicating these data with minimum delays and no corruption and securely is a significant ICT challenge; that is why cyber security is a leading topic for future power systems [37].

The second challenge is the wise planning and implementation of what we can call the grid codes evolution to *standardise* the provision of ancillary services by renewable energy. This should consider tailoring the definition of reserve and inertia to versatile nature of the widely accepted frequency support methods. For example, should the TSOs adopt a pre-populated frequency-active power response or should they be granted a limited access to the holistic controls of renewable power plants to achieve power balance? How should the TSO ancillary service market coordination be achieved without curtailing both system stability and renewable power plants finance [38]? In addition, what is the standard definition of a renewable power plant, as it could be a hybrid energy source with energy storage system?

Author details

Ayman Attya[1]* and Jose Luis Dominguez-Garcia[2]

1 Senior Lecturer in Electrical Power in Engineering, Department of Engineering Technology, University of Huddersfield, UK

2 Head of Power Systems, Department of Electrical Engineering (IREC), Spain

*Address all correspondence to: a.attya@hud.ac.uk

IntechOpen

© 2020 The Author(s). Licensee IntechOpen. Distributed under the terms of the Creative Commons Attribution - NonCommercial 4.0 License (https://creativecommons.org/licenses/by-nc/4.0/), which permits use, distribution and reproduction for non-commercial purposes, provided the original is properly cited. BY-NC

References

[1] ENTSO-E. Network Code on Load Frequency Control and Reserves (LFCR). 2015. Available at: https://www.entsoe.eu/major-projects/network-code-development/load-frequency-control-reserves/Pages/default.aspx

[2] Attya ABT, Hartkopf T. Control and quantification of kinetic energy released by wind farms during power system frequency drops. IET Renewable Power Generation. 2013;**7**(3):210-224

[3] Margaris ID, Papathanassiou SA, Hatziargyriou ND, Hansen AD, Sorensen P. Frequency control in autonomous power systems with high wind power penetration. IEEE Transactions on Sustainable Energy. 2012;**3**(2):189-199

[4] Attya ABT, Dominguez-García JL. Insights on the provision of frequency support by wind power and the impact on energy systems. IEEE Transactions on Sustainable Energy. 2018;**9**(2):719-728

[5] Hafiz F, Abdennour A. Optimal use of kinetic energy for the inertial support from variable speed wind turbines. Renewable Energy. 2015;**80**:629-643

[6] Arani MFM, Mohamed YA-RI. Analysis and mitigation of undesirable impacts of implementing frequency support controllers in wind power generation. IEEE Transactions on Energy Conversion. 2016;**31**(1)

[7] Attya AB, Hartkopf T. Wind turbine contribution in frequency drop mitigation-modified operation and estimating released supportive energy. IET Generation, Transmission & Distribution. 2014;**8**(5):862-872

[8] Attya AB, Hartkopf T. Wind turbines support techniques during frequency drops—Energy utilization comparison. AIMS Energy. 2014;**2**(3):260-275

[9] Li S, Haskew TA, Williams KA, Swatloski RP. Control of DFIG wind turbine with direct-current vector control configuration. IEEE Transactions on Sustainable Energy. 2012;**3**(1):1-11

[10] Attya AB, Ademi S, Jovanović M, Anaya-Lara O. Frequency support using doubly fed induction and reluctance wind turbine generators. International Journal of Electrical Power & Energy Systems. 2018;**101**:403-414

[11] Attya AB, Anaya-Lara O. Provision of frequency support by offshore wind farms connected via HVDC links. In: 5th IET International Conference on Renewable Power Generation. 2016

[12] Ma J, Qiu Y, Li Y, Zhang W, Song Z, Thorp JS. Research on the impact of DFIG virtual inertia control on power system small-signal stability considering the phase-locked loop. IEEE Transactions on Power Systems. 2017;**32**(3):2094-2105

[13] Attya AB, Anaya-Lara O, Leithead WE. Novel concept of renewables association with synchronous generation for enhancing the provision of ancillary services. Applied Energy. 2018;**229**:1035-1047

[14] Attya AB, Anaya-Lara O, Ledesma P, Svendsen HG. Fulfilment of grid code obligations by large offshore wind farms clusters connected via HVDC corridors. Energy Procedia. 2016;**94**:20-28

[15] Johnson A. Fault Ride Through ENTSO-E Requirements for Generators–Interpretation. National Grid. 2013. Available at: http://www2.nationalgrid.com/WorkArea/DownloadAsset.aspx?id=17270

[16] EirGrid. EirGrid Grid Code, Version 6.0. Ireland. 2015. Available at: http://www.eirgridgroup.com/site-files/library/EirGrid/GridCodeVersion6.pdf

[17] Network code for requirements for grid connection applicable to all generator. ENTSO-E. 2013

[18] Vidal J, Abad G, Arza J, Aurtenechea S. Single-phase DC crowbar topologies for low voltage ride through fulfillment of high-power doubly fed induction generator-based wind turbines. IEEE Transactions on Energy Conversion. 2013;**28**(3):768-781

[19] Ezzat M, Benbouzid M, Muyeen S, Harnefors L. Low-voltage ride-through techniques for DFIG-based wind turbines: State-of-the-art review and future trends. In: IECON 2013-39th Annual Conference of the IEEE; Industrial Electronics Society, IEEE. 2013. pp. 7681-7686

[20] Domínguez-García JL, Gomis-Bellmunt O, Trilla-Romero L, Junyent-Ferré A. Indirect vector control of a squirrel cage induction generator wind turbine. Computers & Mathematics with Applications. 2012;**64**(2):102-114

[21] Zhang X, Wu Z, Hu M, Li X, Lv G. Coordinated control strategies of VSC-HVDC-based wind power systems for low voltage ride through. Energies. 2015;**8**(7):7224-7242

[22] Mendes VF, Matos FF, Liu SY, Cupertino AF, Pereira HA, De Sousa CV. Low voltage ride-through capability solutions for permanent magnet synchronous wind generators. Energies. 2016;**9**(1):59

[23] Zhang L et al. Application of a novel superconducting fault current limiter in a VSC-HVDC system. IEEE Transactions on Applied Superconductivity. 2017;**27**(4):1-6

[24] Wang L, Jiang P, Wang D. Summary of superconducting fault current limiter technology. Frontiers in Computer Education. 2012:819-825

[25] Attya AB, Comech MP, Omar I. Comprehensive study on fault-ride through and voltage support by wind power generation in AC and DC transmission systems. The Journal of Engineering. Available at: https://digital-library.theiet.org/content/journals/10.1049/joe.2018.9339

[26] Appendix A Vestas General Specification Brochure. In: Port Dover and Nanticoke Wind project Wind Turbine Specifications Report. 2010. Available at: http://www.homepages.ucl.ac.uk/~uceseug/Fluids2/Wind_Turbines/Turbines/V90_GeneralSpecification_3.pdf

[27] ENTSO-E. NETWORK CODE FOR REQUIREMENTS FOR GRID CONNECTION APPLICABLE TO ALL GENERATORS—REQUIREMENTS IN THE CONTEXT OF PRESENT PRACTICES. ENTSO-E, Belgium. 2012. Available at: http://www.acer.europa.eu/Media/News/Documents/120626%20-%20NC%20RfG%20-%20Requirements%20in%20the%20context%20of%20present%20practices.pdf

[28] Heffron W, Phillips RJIT. Effect of modem amplidyne voltage regulator characteristics. 1952. pp. 692-697

[29] Anaya-Lara O, Jenkins N, Ekanayake JB, Cartwright P, Hughes M. Wind Energy Generation: Modelling and Control. John Wiley & Sons; 2011

[30] Domínguez-García JL, Gomis-Bellmunt O, Bianchi FD, Sumper AJR, Reviews SE. Power oscillation damping supported by wind power: A review. 2012;**16**(7):4994-5006

[31] Kundur P. Power System Stability and Control. New York: McGraw-Hill Inc.; 1994

[32] Clark K, Miller NW, Sanchez-Gasca JJ. Modeling of GE wind turbine-generators for grid studies. 2010;**4**:0885-8950

[33] Attya AB, Hartkopf T. Evaluation of wind turbines dynamic model parameters using published manufacturer product data. In: 2012 {IEEE} International Energy Conference and Exhibition ({ENERGYCON}). 2012

[34] Wang N, Johnson KE, Wright AD. Comparison of strategies for enhancing energy capture and reducing loads using LIDAR and feedforward control. IEEE Transactions on Control Systems Technology. 2013;**21**(4):1129-1142

[35] Rebello E, Watson D, Rodgers M. Performance analysis of a 10 MW wind farm in providing secondary frequency regulation: Experimental aspects. IEEE Transactions on Power Systems. 2019;**34**(4):3090-3097

[36] Goksu O, Teodorescu R, Bak CL, Iov F, Kjaer PC. Instability of wind turbine converters during current injection to low voltage grid faults and PLL frequency based stability solution. IEEE Transactions on Power Systems. 2014;**29**(4):1683-1691

[37] Tøndel IA, Foros J, Kilskar SS, Hokstad P, Jaatun MG. Interdependencies and reliability in the combined ICT and power system: An overview of current research. Applied Computing and Informatics. 2018;**14**(1):17-27

[38] Ackermann T et al. Scaling up variable renewable power: The role of grid codes. 2016

Chapter 7

Frequency-Power Control of VSWTG for Improved Frequency Regulation

Asma Aziz and Aman Than Oo

Abstract

With increasing wind energy penetration and impending grid codes, it is important to enable wind-based power plants to provide sensitive frequency response in grids that may experience irregular frequency fluctuations with noise induced. Transient low-frequency deviations are handled by inertial control, while active power frequency response controller is needed for high-frequency control. A frequency processor-based frequency-active power set point controller architecture for variable speed wind turbine generator (VSWTG) is presented in this paper. Grid frequency processor based on moving averaged frequency and dynamic dead-band is tested for two different grid codes. Generated active power set point is provided to a modified torque-pitch control loop in type 3 and type 4 variable speed wind turbine generator generic models. Delay model of hydro system in a single area load frequency control is applied to investigate frequency support from proposed frequency response controller-based VSWTG. Area frequency response along with VSWTG electrical power support is compared with other droop-based VSWTG model to establish the superiority of proposed frequency-active power controller-based VSWTG over other droop-based VSWTG models.

Keywords: frequency response, VSWTG, active power control, frequency grid code, load frequency control

1. Introduction

Deteriorating power quality due to frequency variations can pose risk of severe economic impacts on a big interconnected electrical network like Australian national electricity market (NEM). In case of rare contingency event like that of 2009 blackout in Victoria [1], load shedding at high level is permissible for frequency regulation and electrical system stability, but for control area like South Australia having wind penetration as high as 70% in some of the days or islanding-prone area like Tasmania, regular load shedding is not acceptable due to high-frequency excursions. Frequency regulation is imperative when ratio of highest contingency loss to system size is relatively high. For large interconnected network like NEM, this ratio is low, but for state network like Tasmania or South Australia where islanding probability is more and there has high wind penetration, frequency regulation from participating frequency-responsive generating plants is highly important on a daily basis.

IntechOpen

System frequency is regulated by injecting active power into electric grid through power plants. This control is indispensable for stable operation of grid ensuring continuous adaptation of generation to demand. Under a wide range of ambient conditions, frequency-responsive power plants are expected to provide automatic power variation with frequency deviation within a given time frame and ramp limit. The conventional power plants which remain connected to grid even in case of frequency deviations accomplish active power-frequency control through turbine governors' reaction to nominal synchronous speed deviations and the respective boiler-turbine frequency-responsive controls. Dead-band and speed droop are two significant parameters in conventional turbine governor action during system event [2]. Long-term system frequency regulation fidelity is highly dependent upon implemented dead-band. Governor dead-band is detrimental to minimum frequency variation needed before the governor action is activated. The relative frequency deviation with respect to the relative change in power output defined as speed droop is always positive for stable regulation. Factors like available headroom, maximum-minimum power capacity of generating units, and power set point enabling frequency operating modes directly affect the total frequency-responsive reserve amount. A generator operating at its maximum generating capacity has almost negligible headroom and is therefore unable to provide any frequency-responsive operation irrespective of governor droop being enabled.

In order to get an emulated synchronous generator-like response, abovementioned parameters like droop, dead-band, and headroom are needed to be considered in variable speed wind turbine generator (VSWTG) generic modeling for frequency-active power control. The generic models are functional models appropriate for the investigation with lower simulation period for large-scale power systems. These generic models can sufficiently represent all dynamics associated with the impact of active power variations in the time range of 50 ms to 100 min. Based on trends in the connection process globally, doubly fed induction generator-based type 3 and full converter-based type 4 VSWTG are dominating renewable energy market. Different type 3 VSWTG models [3–5] exist in literature studies, but wind turbine and wind power plant (WPP) model still lack standardization, and works are in progress. Various software like PSLF, PSS-E, DlgSILENT, etc., have been used for implementing these models. An official version of second generation WECC models was released in 2014, but their adoption has not been up to the desired level as not every type of VSWTG model can be applied in every study [6]. Even though VSWTG model characterization development has been fundamentally accomplished, development of plant controllers especially regarding frequency response control is still under developing stage.

The impending system inertia reduction with increased integration of power electronics-based wind power plants has led transmission system operators (TSO) to establish new grid codes for frequency-based ancillary services from wind farms in big interconnected electrical grid. Grid codes like that of the UK [7] require a response from wind farms under normal conditions and limited up/down response under high-frequency conditions. In case of TSO commanded restriction on active power output, Irish grid code implies all grid-integrated generating units to be capable of operating at a reduced output level. According to Australia's national electricity market (NEM), it is compulsory for generators ≥30 MW to participate in frequency control ancillary services [8]. Operation of the VSWTGs in maximum power tracking mode results in zero spinning reserve for utilization in frequency regulation. In view of compulsory grid code frequency support, this power electronics-based VSWTG requires modified active power control algorithm to support regulation services. Frequency-active power control model is an auxiliary control algorithm implemented in individual wind turbine generator control loop for providing controllable power reserve on demand in the form of spinning reserve or power ramp rate limit to respond to system frequency deviations. Wind turbine active power frequency regulation analysis has received considerable research focus in the form of inertial control and primary frequency control analysis either at individual wind turbine level or at wind farm level [9–14]. Most of the past researches focus on modifying individual turbine control algorithm for analyzing primary frequency control capabilities from wind turbines. These algorithms focus on either achieving de-loaded performance [15–18] to enhance VSWTG's frequency regulation capability or investigated feasibility of droop control on frequency stability [19–24]. The basic idea behind most of the droop-based studies for emulating primary frequency response and inertial response studies is to add an additional signal, sensitive to frequency to the speed or torque controller which appears as reference power set point in VSWTG model, thereby momentarily increasing the wind turbine output power. However, the output reference power set points from these controls are not grid code compatible.

Power set points are the target for required electrical energy generation from wind plants. These targets can be based on available wind, operational mode, or required frequency response. Active power set point characteristics and the expected behavior of the different generating units under different set points are defined in country-specific grid codes. Centralized wind farm control for active power set point distribution to individual turbines from wind farm controller was presented in [25]. In this model, wind turbines receive power set point distributed through a centralized wind farm controller which is provided by transmission system operator, and wind turbines did not individually respond to frequency excursions at their terminals. Another research study for active power control of wind turbine presented a control system for tracking a power reference and provides a primary frequency response with constant droop in the absence of any dead-band with diminishing turbine structure loading [26]. [27] describes three de-rating command mode for wind turbine operation as $DR_{cmd}P_{rated}$, $P_{avail} - (1 - DR_{cmd})P_{rated}$, and $DR_{cmd}P_{avail}$, where DR_{cmd} is the de-rating command set point, P_{rated} is rated output of wind plant, and P_{avail} is the available wind power. In another study, three different operating modes (de-rated, absolute spinning reserve, and relative spinning reserve) [28] are defined. As an example, operating mode 1 is defined as de-rated or normal such that:

$$\boldsymbol{P}_{ref} = \left\{ \begin{matrix} \boldsymbol{P}_{operator} & \forall \boldsymbol{P}_{operator} & \leq \mathrm{P}_{avail} \\ \boldsymbol{P}_{avail} & \forall \boldsymbol{P}_{avail} & \leq \boldsymbol{P}_{operator} \end{matrix} \right\}$$

In the view of above discussions, the objectives of the current paper are as follows:

1. To present a grid frequency processor scheme based on dynamic dead-band around moving averaged frequency instead of conventional static dead-band for generation of continuously varying, frequency-sensitive active power reference set point for VSWTG and its corresponding effect on system primary frequency control.

2. Frequency-sensitive response and frequency limited sensitive response are explored through dynamic dead-band concept and nonsymmetrical droop for power variation at wind turbine level.

3. Grid code-sensitive frequency active power controller-based type 3 and type 4 wind turbine response is analyzed with response of frequency-dependent linear droop controller-based wind turbine generator model.

4. Frequency controllers' performance is compared with and without wind power in algorithm.

This paper is structured in the following sections. Section 2 explains the modeling of grid code-sensitive frequency controller for active power response in VSWTG. Individual blocks including grid frequency processor highlighting effect of variation of dead-band and wind turbine control loop are explained. Section 3 discusses test system load frequency control model incorporating hydro power plant and frequency-responsive power wind power plant. MATLAB-based simulation results are used to compare frequency response controller I- and frequency response controller II-based VSWTG response.

2. Grid code-sensitive frequency controller for active power response in VSWTG

2.1 Dead-band: static vs. dynamic

Generator droop characteristic means that generating unit will inversely change power output level in proportion to system frequency. Responding to frequency deviations, generating units depart from their dispatch targets according to the set droop characteristics. This droop behavior can be avoided by implementing a dead-band on each unit. Static dead-band is thus a symmetrical buffer zone on either side of 50/60 Hz frequency that compensates for frequency noise. Droop response of generating machine from controlling power output of machine is avoided when frequency lies within the upper and lower dead-band range. Dynamic dead-band incorporates a buffer zone on either side of moving averaged frequency signal instead of nominal frequency signal. Both these dead-bands are represented in **Figure 1**.

2.2 Grid frequency processor background

Providing grid frequency directly to wind turbine controller will generate noise-induced frequency-sensitive active power set point. This will result in noisy output power, thus providing frequency response for any frequency error even due to grid noise. Noisy power output will have an adverse effect on supplied power quality

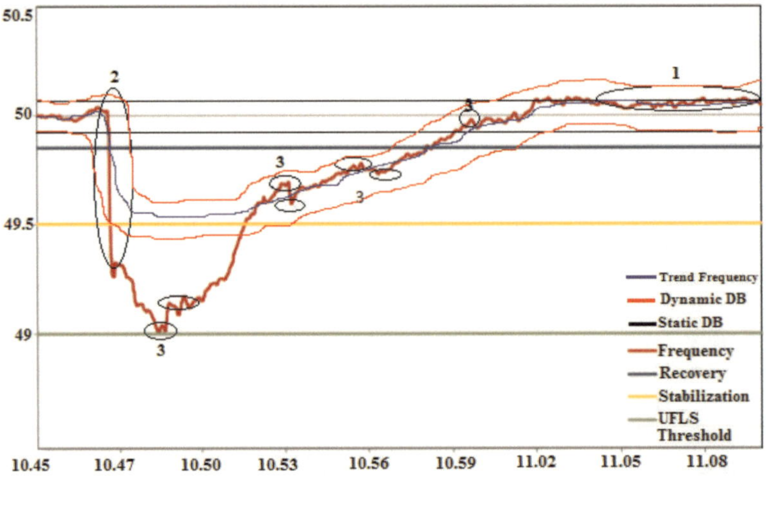

Figure 1.
NEM mainland frequency 1045–1109 h during a contingency event [1].

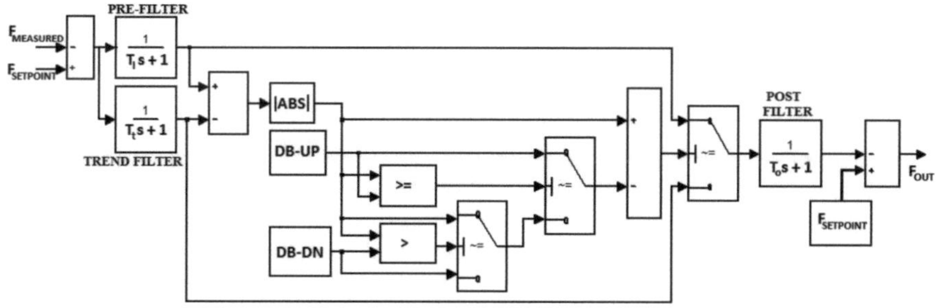

Figure 2.
Basic structure of grid frequency processor.

and turbine life. An analysis of electrical grid dynamics can serve as the base of grid processing system for distinguishing different types of frequency signals. **Figure 2** shows NEM mainland real frequency trace during 2009 contingency event when 3205 MW of total generation disconnected automatically, resulting in under frequency load shedding across the Australian NEM interconnected system [1].

Though we could not get enough data points to represent a very clear picture here, but still three different frequency signals 1, 2, and 3 as defined below can be distinguished in **Figure 1**. Electrical grid frequency signal can be considered as composition of three types of signals [29]:

1. Low ramping signals arising due to normal trend of generation-demand and grid dynamics over long term of several minutes

2. High-amplitude, high-frequency range during contingency which is normally low in occurrence

3. Low-amplitude signals in high-frequency range due to stochastic grid noise with high occurrence rate

First two signals have very low effect on turbine life, while highly occurring type 3 noise signals have ample effect on turbine life time and should be suppressed. A closer look at **Figure 1** shows that trend frequency signals 1 are usually around the static dead-band, while high-amplitude signal 2 is always outside the static dead-band. Discerning these signals just on the basis of frequency through filters is not a viable option as they are present in whole frequency signal over a time period. Implementing a dynamic dead-band centered around the trend frequency signal in a grid processing system would distinguish these types of signals.

2.3 Frequency processor model

A grid frequency processing system centered around moving averaged frequency signal was proposed in [29]. A modified version of the frequency processor is implemented in current study for utilization in active power set point controller for variable speed wind turbine generator system operating under frequency-responsive mode. Basic structure of grid frequency processor block which provides dynamic dead-band-based processed frequency output is shown in **Figure 2**. Frequency processor takes grid frequency at PCC and nominal frequency (50 or 60 Hz) as input signal. Three low-pass filters and dead-band constitute the processor model. Moving average of the measured grid frequency is termed as a trend frequency which characterizes long-term behavior of grid frequency. The theoretical moving average filter (MAF) is mathematically expressed as

$$y[i] = \frac{1}{M} \sum_{j=0}^{M-1} x[i+j] \tag{1}$$

where x is the input frequency signal, y is the output frequency signal, and M is the number of points in average [30]. A moving average filter will cause minimal change to a signal whose period is long compared to the filter window length, because the filter's window only "sees" a small and relatively constant part of the oscillating signal at each moment. MAF, which can be practically implemented as the finite impulse response filter (FIR), produces lowest noise component in output signal by equally treating all incoming signal.

A simple moving average filter acts as a low-pass filter. A low-pass filter passes very low frequencies with minimal change, but it reduces the amplitude of high-frequency signals or of high-frequency components in a complex signal. A low-pass filter with a cutoff of $f_{CO} = 0.443/T_t$ has been used to act as MAF to construct trend frequency in this study. Trend frequency is assumed to lie within the dead-band, so selecting the trend filter time constant between 8 and 30 s will provide a cutoff frequency 50$\pm$0.015 to 50$\pm$0.05 Hz. The time constant for prefilter is taken as 0.5 s which gives a cutoff frequency of 50$\pm$0.89 Hz. This time constant will ensure filtering under normal frequency variation just before the under-frequency load shedding (UFLS) as shown in **Figure 1**. Trend frequency signal is subtracted from prefiltered measured frequency, and resulting absolute frequency difference is then passed through dead-band algorithm. The dead-band algorithm dynamically limits the range of the input signal according to the upper and lower dead-band limits. If DB-UP $\leq |F_{Filter} - F_{trend}| >$ DB-DN, output is set to zero. If $|F_{Filter} - F_{trend}| >$ DB-UP, the output appears as the input shifted down by the DB-UP. If $|F_{Filter} - F_{trend}| <$ the DB-DN, output appears as input signal shifted down by the DB-DN, as indicated in **Figure 3**.

Flowchart for basic frequency processor algorithm is shown in **Figure 3**. **Figure 4** shows the measured grid frequency and processed out frequency for different threshold values for selector switch. These threshold values are selected based on different upper and lower dead-band limits. For example, for a system with dead-band limit of $\pm$0.03 Hz, threshold is selected as 0.03. In this study, an upper dead-band of 0.015 Hz which is the standard value for most of the TSO [31, 32] and lower dead-band of 0.001 Hz are selected, so 0 is selected as the threshold value. As per NERC policy, total dead-band applied should be limited to 0.035 [33].

Output processed frequency is basically composition of trend frequency and measured filtered frequency as seen in **Figure 5**. We can notice output processed frequency in black color following trend frequency in green color most of the time when dead-band output is zero, while processed frequency follows measured frequency in red color whenever threshold increases above zero. As stated previously, frequency processor is dependent upon implemented threshold value and dead-band limits which can be set as per TSO requirements.

A 14-generator NEM model is a simplified model of the eastern and southern 50 Hz Australian electrical networks, which was originally proposed for small-signal stability studies [34]. In the original model, there are 14 generators, 5 static VAR compensators (SVCs), 59 busses, and 104 lines with voltage levels ranging from 15 to 500 kV. It is assumed that all thermal and hydro power plants have a standard steam turbine governor (i.e., IEEEG1) and hydro turbine governor (i.e., HYGOV), respectively. **Figure 6** presents the grid frequency from one of the busses from NEM model which is provided to grid frequency processor. The resulting processed and filtered frequency is shown in **Figure 7**. The next section discusses the application of grid frequency processor in different types of frequency controllers for generating frequency-sensitive power set points.

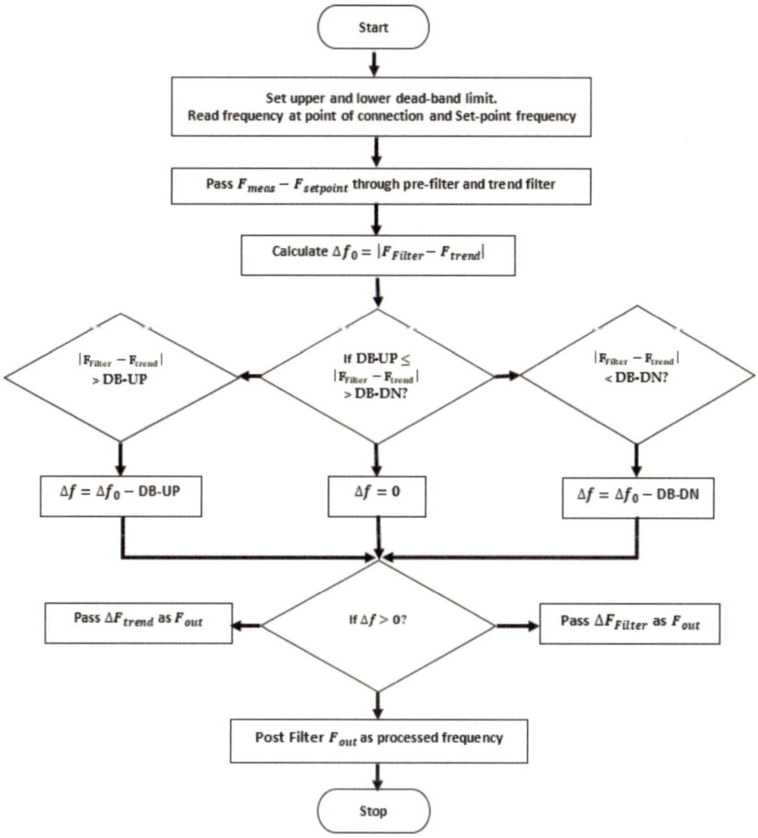

Figure 3.
Algorithm for grid frequency processor.

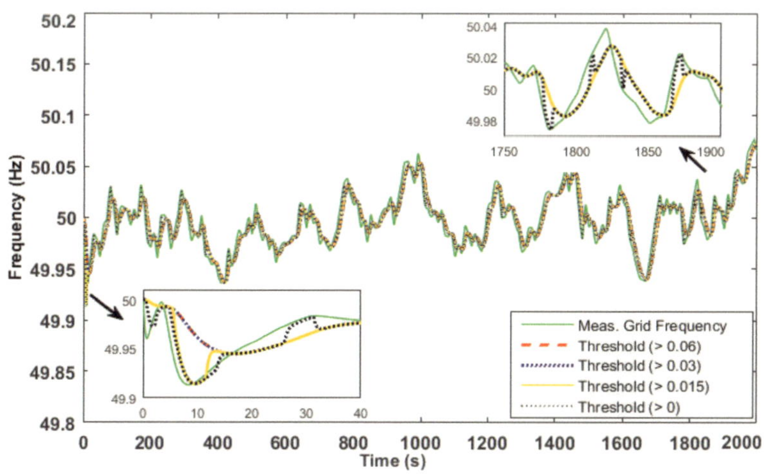

Figure 4.
Grid frequency vs. processed frequency at different threshold values (presented in different colors).

2.4 Frequency response controller model-I

Frequency response controller model-I as shown in **Figure 8** incorporates basic frequency processor in its structure to provide frequency response operation. This frequency controller block can provide two types of frequency response according

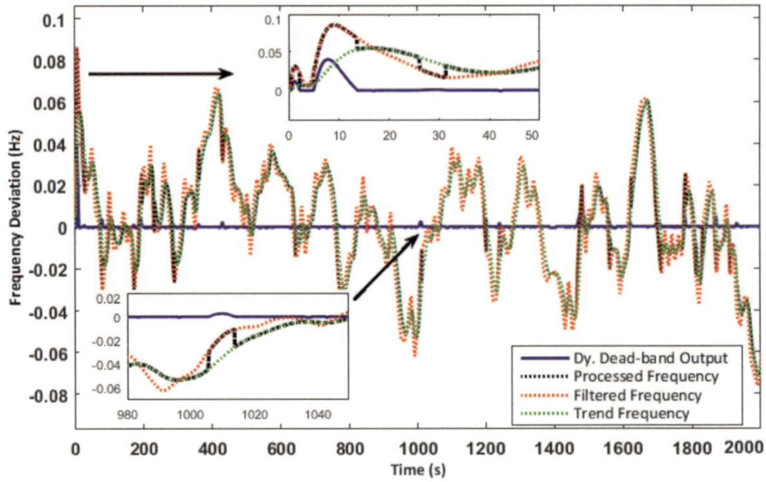

Figure 5.
Processed frequency, trend filter output, low-pass freq. Filter output, and DDB output when threshold applied is >0.

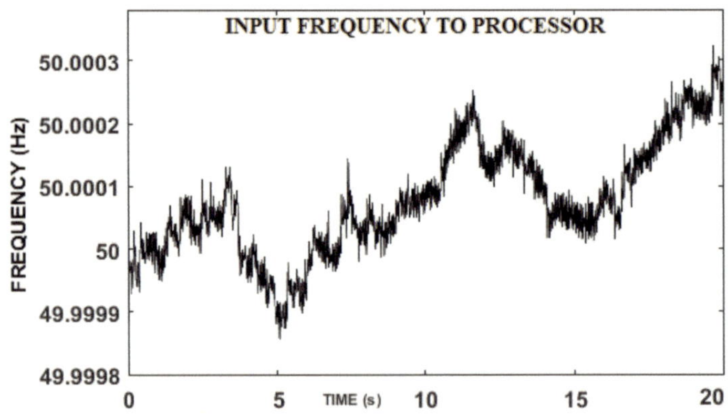

Figure 6.
An example of input frequency obtained from NEM 14-generator model provided as input to frequency processor.

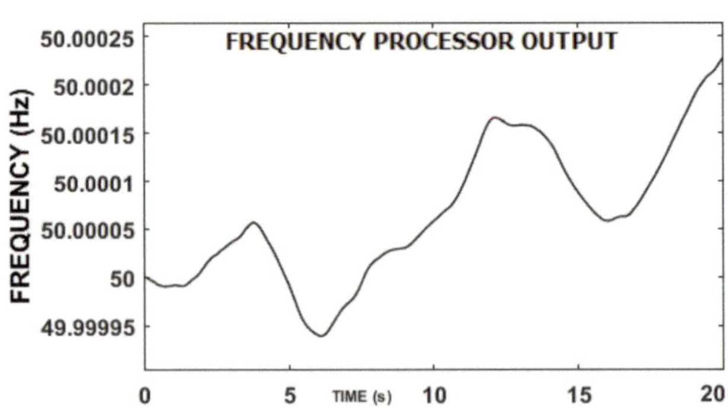

Figure 7.
NEM 14-generator model processed frequency obtained from grid frequency processor.

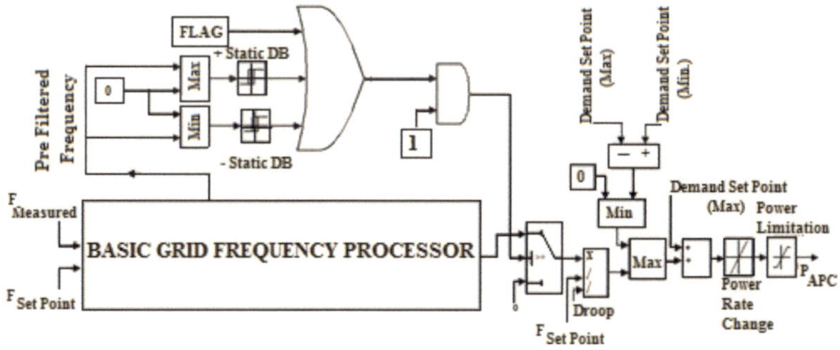

Figure 8.
Frequency response controller model-I.

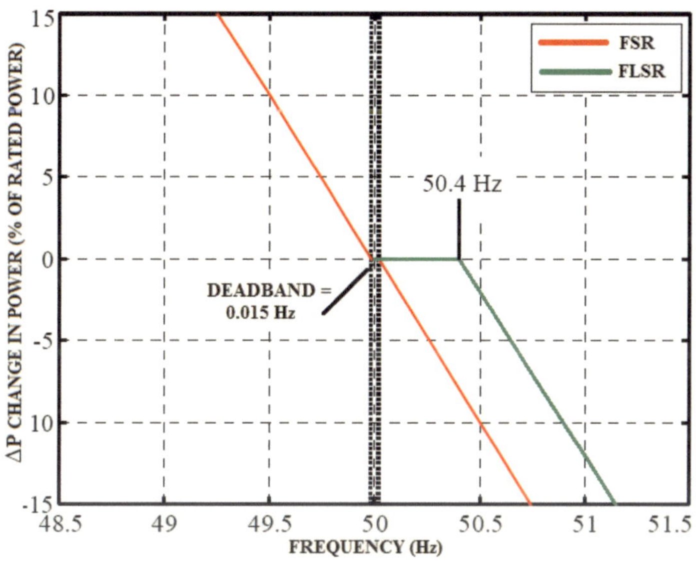

Figure 9.
Frequency grid code UK [35].

to the grid code as shown in **Figure 9**. Algorithm of this frequency response controller is explained through the flowchart given in **Figure 10**.

2.4.1 Frequency-sensitive response mode

By setting FLAG equal to 1, frequency-sensitive response mode can be activated. Power set point will change proportionally to both up and down frequency deviation from reference frequency signal. Processed output frequency is generated through the coordination of the trend control frequency signal and dynamic dead-band. If measured frequency follows trend control frequency signal within dynamic dead-band, trend frequency signal is forwarded for generating droop power response; otherwise measured frequency signal is forwarded. VSWTG final active power set point is generated through an algorithm implemented on frequency response power and power demand set point.

2.4.2 Frequency limited sensitive mode

By setting FLAG equal to 0, limited frequency-sensitive response mode can be activated to provide high-frequency response. There should not be any power

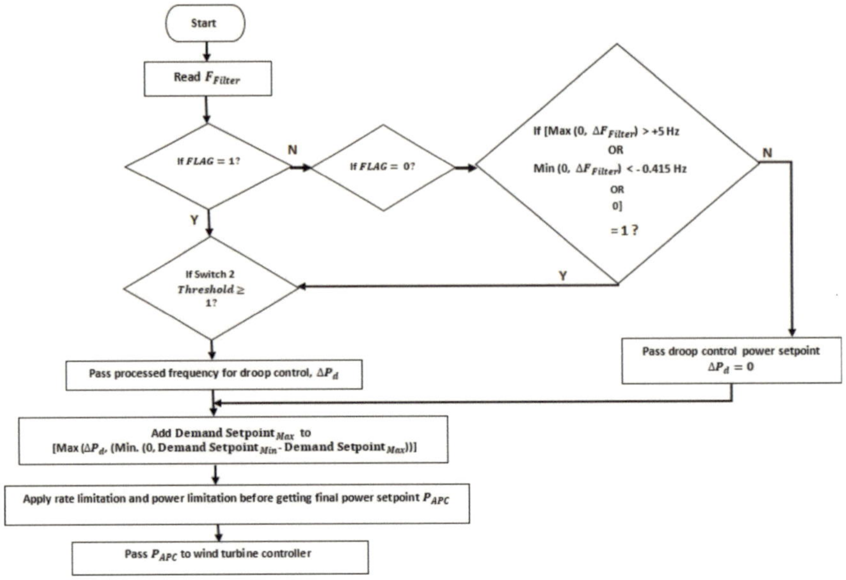

Figure 10.
Algorithm for frequency controller-I FSR and FLSR response.

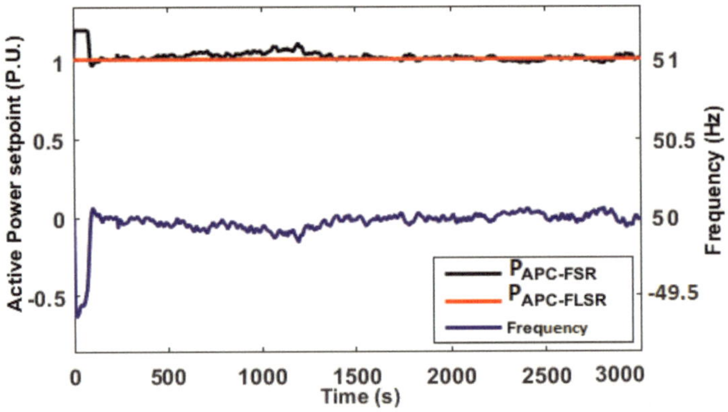

Figure 11.
Example of frequency controller-I FSR and FLSR response.

variation corresponding to the frequency changes except when the frequency exceeds the upper limit. Insensitive mode incorporates a static dead-band around the frequency set point. If frequency error remains within static band, trend frequency signal and dynamic dead-band has no influence on output power.

Frequency response operation of the controller-I can be understood by **Figures 11** and **12**. As shown in **Figure 11**, when working under frequency-sensitive mode, wind frequency response controller-I provides high active power set point in case of low system frequency and low active power set point in case of high frequency. In limited frequency response mode, frequency response controller-I provides only low active power set point when system frequency is more than 50.4 Hz for reduced power generation. In case of frequency being lower than 50.4 Hz, wind turbine follows the demand set point as shown in **Figure 12**. This type of frequency controller is highly compatible with grid codes like that of the UK and Australia which requires full response from wind farms under normal conditions and limited up/down response under high-frequency conditions.

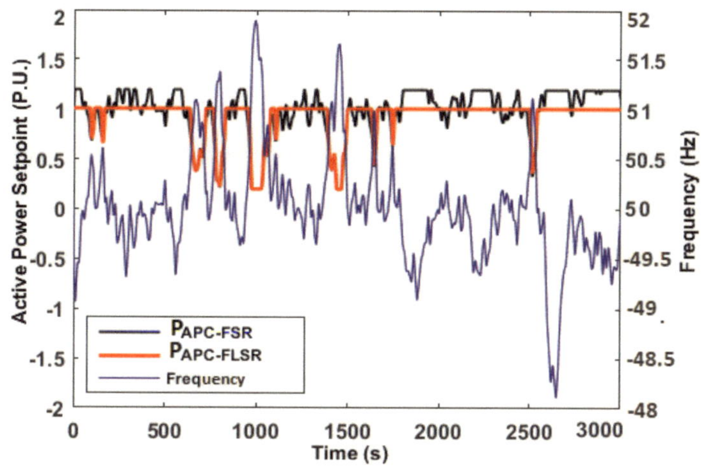

Figure 12.
Another example of active power set point generated when employing frequency response controller-I in two different modes.

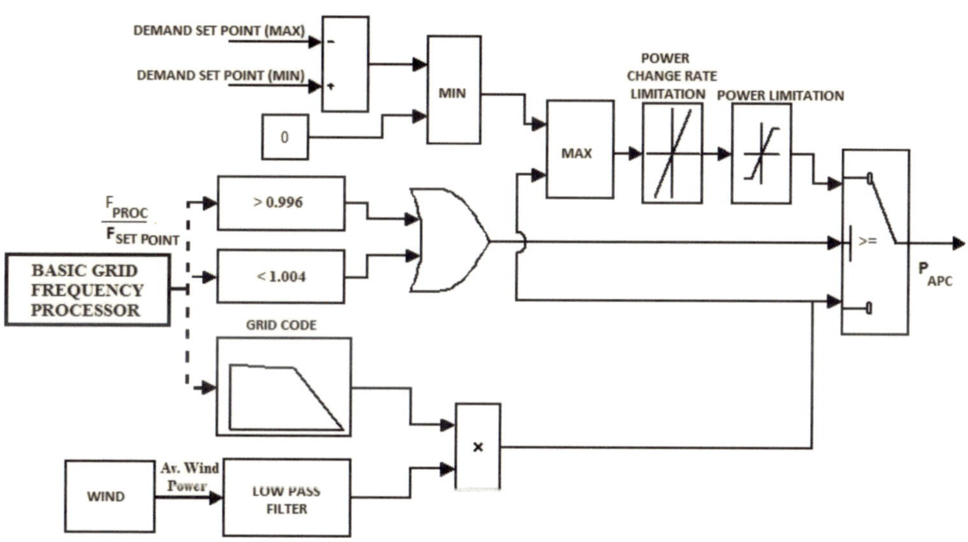

Figure 13.
Frequency response controller model-II.

2.5 Frequency response controller model-II

Frequency control model as shown in **Figure 13** implements the setting of active power set point according to typical nonsymmetrical droop curve very similar to Irish grid code as shown in **Figure 14**. Wind speed and processed frequency from basic frequency processor block are its two inputs. A wind profile module is implemented to provide variable wind speed as given in **Figure 15** for this study. The available wind power in per unit (p.u.) is calculated as a function of wind speed:

$$P_{avail} = \frac{1}{2}\rho A V^3 \tag{2}$$

where P_{avail} is the wind power [W], ρ is the air density [kg/m^3], v is the wind speed [m/s], and A is the swept area [m^2] of rotor disk that is perpendicular to the

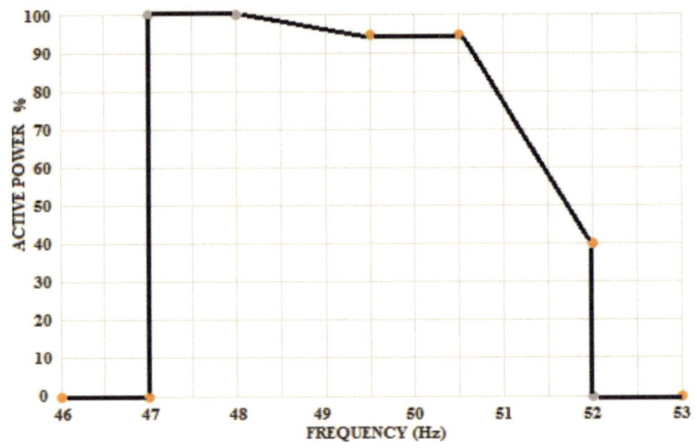

Figure 14.
Nonsymmetrical droop curve for frequency power regulation.

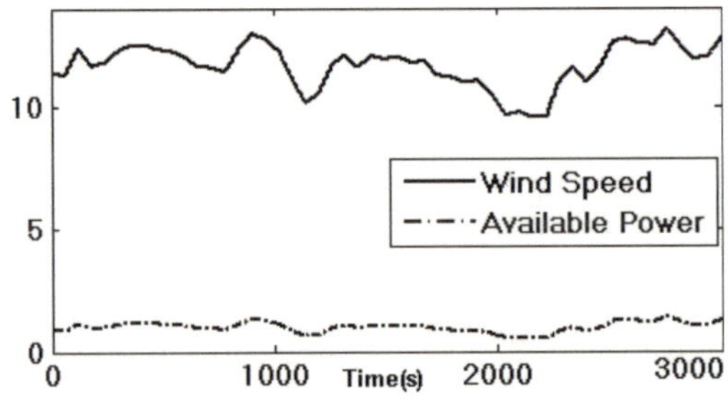

Figure 15.
Variable wind speed (m/s) and available wind power (p.u.) as applied in this study.

wind flow [36]. Filtered available wind power is then multiplied with the power set point received from algorithms defining respective grid codes for incoming frequency. Power set point is then compared with maximum and minimum power restrictions to generate final active power set point. Limited set point is achieved when additional condition is imposed as IF frequency variation <0.996 AND frequency variation >1.004, $P_{set-point}$ EQUALS P_{avail} x $P_{Grid-code}$. Implementation of this restriction helps in maintaining maximum output from wind turbine in case of limited deviation. Flowchart of this controller is given in **Figure 16**.

A wind power plant is running as spinning reserve will produce less power at all wind speeds, thereby always providing a power reserve. Grid code controller will act as spinning reserve controller to vary the WPP grid power production as per transmission system operator request. In the case of wind farm operation, turbine set point can be multiplied by active power demand provided by farm controller to generated frequency-responsive power demand. **Figure 17** shows various power set points generated through grid code compatible frequency controller-II at variable wind speed and fluctuating grid frequency. Black color is the power set point required according to an example grid code, while green is the available wind power. This available power is multiplied with grid code power and limited, thereby providing a reserve power to be used in frequency deviations. We can notice that VSWTG electrical power output follows the limited active power set point provided from the frequency controller.

2.6 Variable speed wind turbine generator generic model

Type 3 and type 4 variable speed wind turbine generator VSWTG generic models adopted from [37] are applied in studying frequency-responsive active

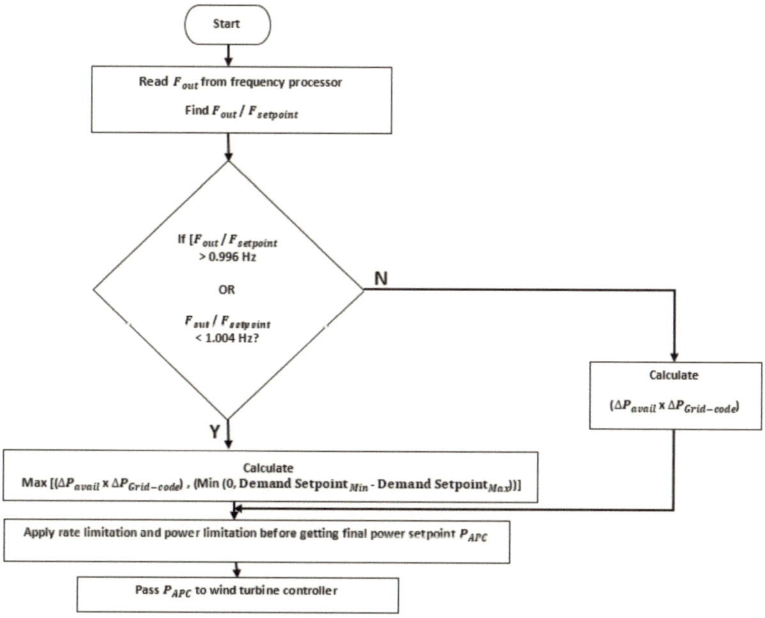

Figure 16.
Flowchart depicting algorithm for frequency response controller model-II.

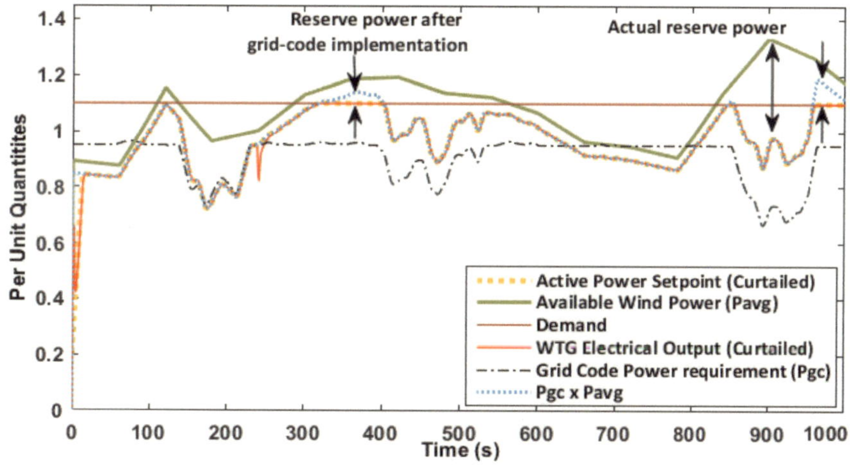

Figure 17.
VSWTG power set points with limited power set point as grid controller output set point and corresponding generated power.

power control. Generic models of both type 3 and type 4 VSWTG have three basic blocks as shown in **Figure 18**: turbine model, generator/converter model, and electrical control model. Detailed model of these blocks can be obtained from [37, 38]. With all similar components and parameters, the major distinguishing factors between type 3 and type 4 VSWTG are in terms of electrical control model. Type 3 electrical control model is represented by flux and active current command, while type 4 electrical control model generates a reactive current command also along with active current command and includes a dynamic braking resistor and converter current limit. During frequency response, real power has priority, so detailed reactive power loop is not applied in both models in this study. The turbine control system in both type 3 and type 4 frequency-responsive models has the common

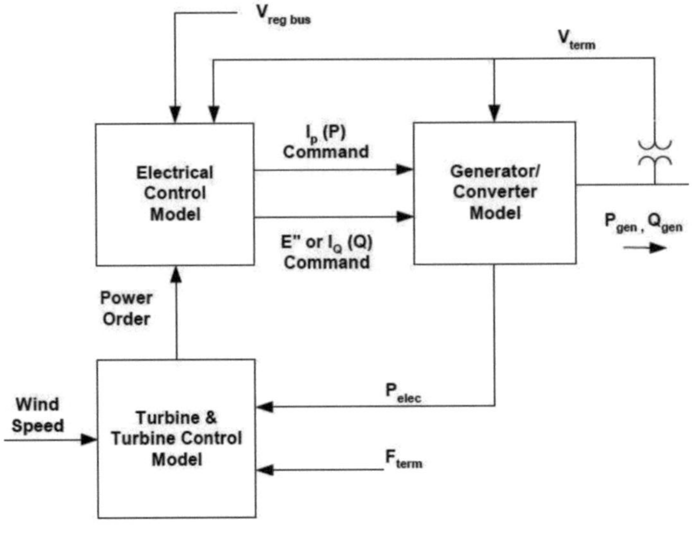

Figure 18.
Modular model for variable speed wind turbines.

objective of controlling power production while maintaining rotor speed below minimum limit. Generic turbine control involves two control loops which receive speed error as input and deliver two control outputs: (1) wind turbine reference power order provided to the converter electrical control and (2) pitch reference value to pitch controller. Rotor speed is controlled as per power command through torque limitation in speed control loop. The current study applies a modified wind turbine control loop which is shown in a red dotted line. One mass lumped mechanical shaft model and detailed aerodynamic model as given in [38] are applied in these models (**Figure 19**).

Power order from torque controller is altered by passing grid code frequency-responsive active power set point P_{APC} through power response rate limiting block. Different gain values are applied to torque controller in type 3 and type 4 VSWTG and provided in the Appendix. Rotor speed error is given as input to both these controls. Final power order ($P_{final-ord}$) is generated by adding grid code limited active power set point from frequency response controller block to difference between active power limited set point power and power order from speed controller ($P_{APC} - P_{inp}$) and can be represented as

$$P_{final-ord} = P_{APC} + \left(\frac{sT_w}{1+sT_w}\right)(P_{APC} - P_{inp}). \quad (3)$$

Ramp rate is implemented by including a washout filter whose time constant (T_w) is detrimental to rate limit imposed on changes in power order. Wind turbine regulates the electrical power according to frequency-responsive final power order ($P_{final-ord}$). A combined torque-pitch control method utilizing P_{APC} is applied in this study to obtain reserve power mode operation. Pitch compensation block provides the necessary margin for frequency-responsive option in current study. Pitch compensation block takes (P_{APC}) generated from frequency-responsive controller instead of rated 1 p.u. reference power. Mechanical power and corresponding shaft speed of wind turbine is controlled through pitch controller and pitch compensation loop. Pitch controller enables control of aerodynamic wind power by rotor blade pitching in order to regulate turbine torque. Maximum power of 1.2 p.u.

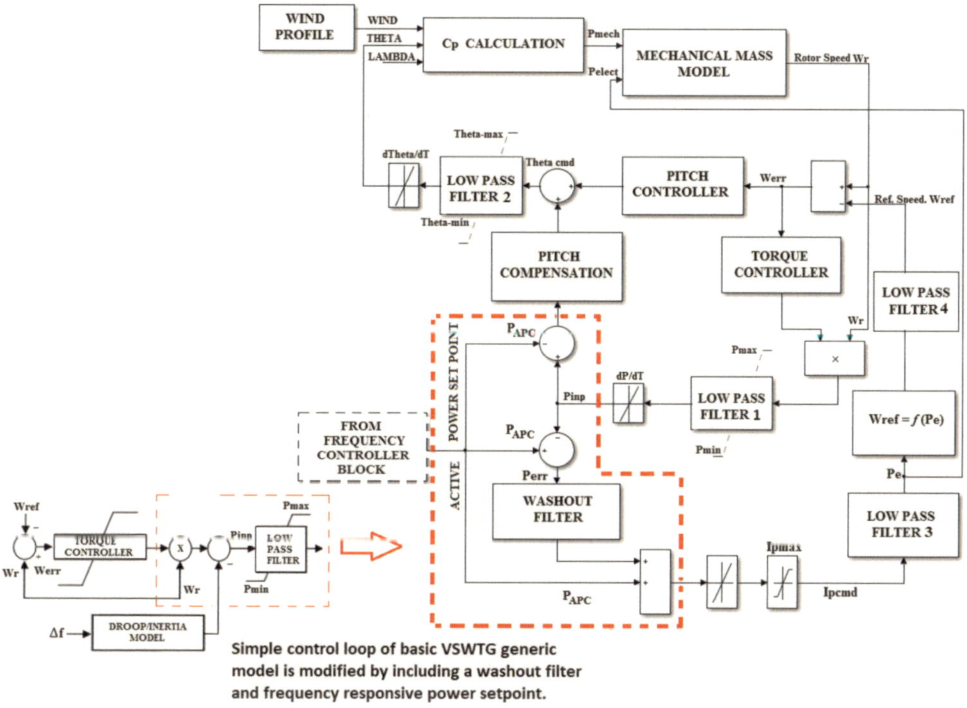

Figure 19.
Type 3 VSWTG model with active power controller feeding power set point PAPC to power controller loop, pitch controller loop, and speed controller loop.

is available at zero pitch angle, while it is highly reduced with the highest pitch angle. Pitch angle is obtained through a series of proportional integral (PI) regulator and mathematically expressed by the following equations:

$$\theta_{cmd} = \frac{d(K_{ip}(\omega - \omega_{ref.}))}{dt} + \frac{d(K_{ic}(P_{max} - P_{set}))}{dt} + K_{pp}(\omega - \omega_{ref}) + K_{pc}(P_{max} - P_{APC}) \tag{4}$$

If $0.15\ \text{p.u.} \le P_{elect} \le 0.75\ \text{p.u.}$, Then $\omega_{ref} = -0.79131P^2_{elect} + 1.526046P_{elect} + 0.49188$.

Else If $P_{elect} \ge 0.75$ p.u., Then $\omega_{ref} = 1.2$ p.u.

Else If $P_{elect} \le 0.15$, $\omega_{ref} = 0.689$ p.u.

More details about the model and components can be referred from [37, 38] (**Figure 20**).

In maximum power point tracking (MPPT) operating mode, turbine power set point is determined, such that

$$P_e = K_{opt}\omega_r^3 \text{ and } K_{opt} = \frac{0.5\pi R^5\rho}{\lambda_{opt}{}^3} C_p(\lambda_{opt}, \beta_0). \tag{5}$$

During maximum power point tracking operational mode, wind turbine electrical power output P_e is equal to MPPT power set point P_{opt}, rotor speed ω_{VSWTG} is equal to ω_{opt}, and pitch angle $\beta = \beta_0 = 0°$. When turbine switches from MPPT mode to commanded power mode, power set point P_{opt} changes to $P_{final-order} = P_e$. VSWTG power drops below mechanical power, thereby increasing rotor speed to

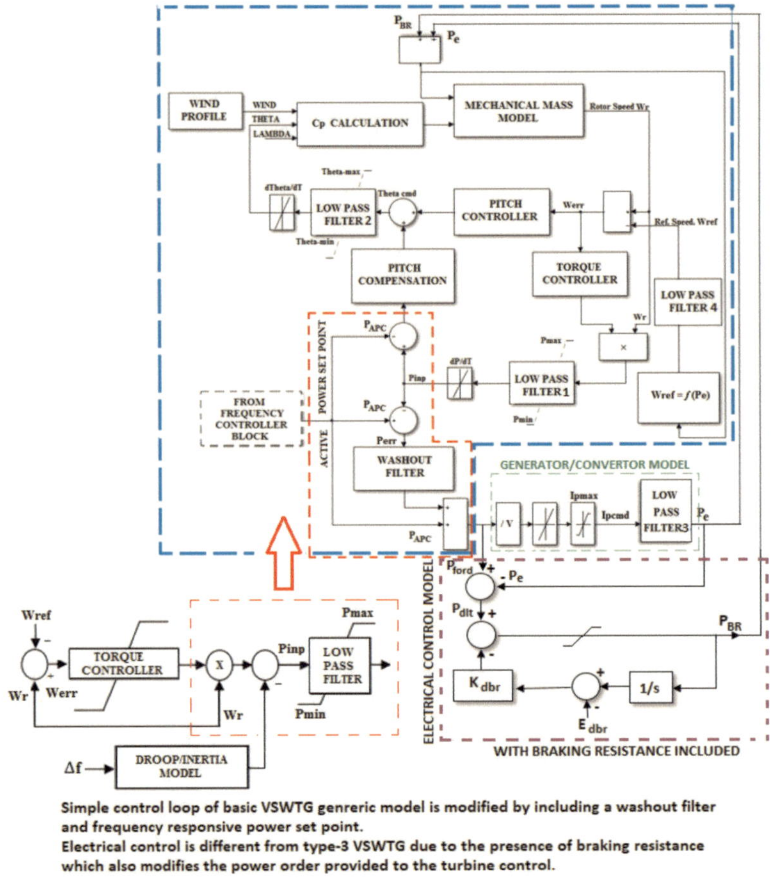

Figure 20.
Type 4 VSWTG model with active power controller feeding power set point PAPC to power controller loop, pitch controller loop, and speed controller loop.

W_r. This increase in rotor speed is controlled by increasing pitch angle β to β_1 which is provided through pitch controller in coordination with pitch compensator. Consequently, power-speed characteristic of wind turbine is varied which is shown in red in **Figure 21**, and rotor speed settles at ω_{max}. To guarantee a safe switching between MPPT mode and commanded mode and safeguard turbine operation

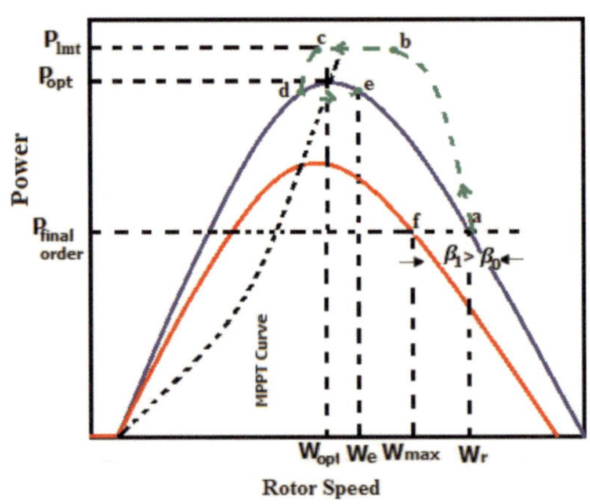

Figure 21.
Characteristics curve of type 3 VSWTG under MPPT and reserve power mode.

beyond rated values, final power order is limited to the optimum power value ($P_{opt} = K_{opt}\omega_r^3$). The turbine will continue to run in MPPT mode unless there is a change in wind speed or change in active power set point from frequency controller. Active power set point from frequency controllers is limited as per desired reserve power of 10% in this study.

In reserve power mode operation, assume steady state condition represented by point a for blue curve at some wind speed such that turbine mechanical power equal to VSWTG electrical power $P_e P_{final-order}$. In the case of frequency deviation due to generation loss, frequency-responsive VSWTG's power will increase to compensate for this loss and may reach VSWTG rated capacity limit shown by point b in **Figure 21**. Power extraction due to kinetic energy will continue until rotor speed hits the minimum limit at point d. After this point, the mechanical power will be more than the electrical power, and rotor will again speed up to finally settle at point e with speed W_e. All interconnected grid code-compatible VSWTGs are required to ensure a fast pitch control as well as other mechanical controller to be able to participate in a frequency response services in an acceptable time.

2.7 Test system: single area load frequency control (LFC) model

Wind penetration in low inertia power systems or system with low primary frequency response has an adverse effect on frequency stability. In order to explore the frequency-responsive wind power plant footprints on the electrical grid frequency regulation performance, a single area load frequency control (LFC) system model using MATLAB/Simulink software is analyzed. Control area power system model as shown in **Figure 22** incorporates hydro power plant along with frequency-responsive wind farm. Detailed modeling of hydro power plant is given in [2]. Since the aim of current study is to assess the improvement provided by frequency-responsive VSWTG model, only a brief description on load frequency modeling is provided in this section. In conventional LFC system, frequency and active power output of generating plants including wind power plants in a single control area is related as

$$\Delta F(s) = [\Delta P_{Gen}(s) + \Delta P_{Wind}(s) - \Delta P_{Load}(s)]\frac{K_P}{1 + ST_P} \tag{6}$$

where $T_P = \left(\frac{2 * H_{eq}}{D * f^0}\right)$ is defined as the power system time constant and $K_P = \frac{1}{D}$ is the power system gain.

LFC operation is accomplished through integral controller, implementing flat frequency control which implies area control error (ACE) as change in frequency Δf.

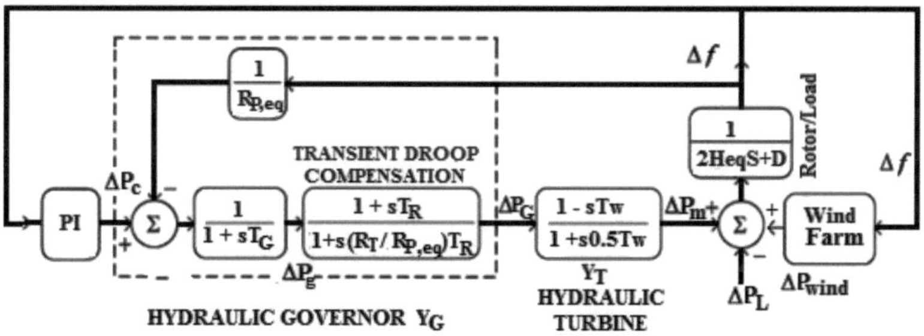

Figure 22.
Single area hydro governor turbine model [2] with integrated wind farm.

$$ACE_i = \Delta f \tag{7}$$

$$\Delta P_{ci}(t) = -K_i \int (ACE_i)dt. \tag{8}$$

Changed droop setting according to wind penetration is given as

$$R_{new} = \frac{R_{old}}{1 - L_p}. \tag{9}$$

On a system-base value of 5000 MW and assuming 0.05 p.u. extra support from VSWTG during frequency excursions, wind-integrated system modified inertia constant, H_{eq}, is calculated as [16]

$$H_{eq} = H_{system}(1 - L_p) + H_{WT}L_p \tag{10}$$

$H_{system} = \frac{\sum_i^N S_{ni} * H_i}{S_{ystem}}$, where $S_{system} = \sum_i^N S_{ni}$

where H_i and S_i are the inertia rating and apparent power rating of individual generating units and L_p is the wind penetration level. For frequency-responsive wind plant, wind inertia contribution H_{WT} to power system by providing ΔP_e extra active power when system is subjected to step load disturbance ΔP_L is calculated as

$$H_{WT} = \frac{\frac{-T_d D}{lnX} - 2H_{eq}(1 - L_p)}{2L_p}; \tag{11}$$

$$X = e^{\frac{-T_d D}{2H_{eq}(1-L_p)}}\left(1 - \frac{\Delta P_e L_p}{\Delta P_L}\right) + \frac{\Delta P_e L_p}{\Delta P_L}. \tag{12}$$

Total time delay T_d associated with hydro governor turbine model is calculated on the basis of delay theory originally given in [39]. Summation of governor time constant, valve motion delay time, and turbine response time delay results in total time delay at which minimum frequency deviation occurs after system disturbance. Value of T_d is calculated as 3.7814 s for hydro governor-turbine model with system parameters given in [16]. System simulations are performed and compared for LFC model with similar parameters but integrated with five different frequency-responsive wind plant model. A comparative study is made for frequency response and its indicators, effect on VSWTG electrical support, and corresponding rotor speed.

The following points have been implemented in this simulation study:

1. A constant load disturbance is applied.

2. All VSWTG models are provided with the same wind speed as shown in **Figure 15**.

3. Frequency response controller-I and frequency response controller-II are implemented for VSWTG model which are shown in **Figures 19** and **20**.

4. Same droop setting of 0.0315 p.u. MW/p.u. Hz is used for all individual generators including hydro and wind plants.

5. Equivalent droop setting of area 1 changes with wind penetration. A 10% wind penetration is used in simulation.

6. The same system inertia is applied in control area model.

All simulation parameters including LFC system parameters are provided in the Appendix.

3. Results and discussions

3.1 Comparison of frequency controller response

Frequency-sensitive response is studied for step load change and with/without considering available wind power in power set point generation algorithm. In **Figure 23a**, **b**, it can be noticed that for any frequency variation outside the dead-band limit, there is a change in active power set point, and VSWTG tracks this power set point with some delay. When available wind power is not considered in algorithm, output power set point tracks demand set point. To investigate the effect of the available wind on frequency controller-I FSR output, power set point algorithm is modified such that

$$\text{If } \left(\frac{F_{Grid}}{F_{set-point}} > 0.996\right) \text{ AND } \left(\frac{F_{Grid}}{F_{set-point}} < 1.004\right), P_{set-point} \text{ EQUALS } P_{avail-wind},$$

else power set point is set as per the algorithm shown in flowchart.

In can be seen in **Figure 23b** that power set point tracks available wind power during normal frequency variation but changes during frequency disturbance. Its value then depends upon maximum value out of droop power or demand set points. VSWTG output traces the provided power set point, but inclusion of wind power can add to significant delay in VSWTG processing. **Figure 23c** shows the frequency-sensitive limited response where a constant power set point is provided if frequency deviation is within a set limit. Frequency-sensitive limited response is provided in the form of power set point variation only when the frequency deviation is more than the set limit: 50.4 in current study. **Figure 23e** shows the frequency-sensitive limited response when frequency increases above 50.4 Hz in the form of decreased power set point from frequency controller. **Figure 23d** presents the frequency controller-II response. Unlike frequency controller FSR response, where power set point tracks the demand set point, frequency controller-II power set point is highly dependent upon grid code power requirement, available wind power, and demand set points. It can be noticed that under all similar parameters and limitation, VSWTG output during frequency disturbance is reduced when wind power is considered in algorithm.

3.2 Comparison of frequency-sensitive type 3 and type 4 VSWTG response

Control area frequency variation under the effect of frequency controller-based wind plants is analyzed in this section. Type 3 and type 4 VSWTG electrical output and rotor speed are also presented when operating under frequency-sensitive power set point. **Figure 24a** and **b** gives the system frequency when 10% penetration of frequency-sensitive grid code compatible type 3 wind plant is integrated along with hydro plant. Maximum frequency drop is 49.99 Hz for 0.01 p.u. load disturbance, while it increases up to 49.83 Hz for 0.1 p.u. load disturbance. The best frequency response under low load condition is observed for frequency controller-II-based type 3 VSWTG integration, while at higher load disturbance, frequency nadir point

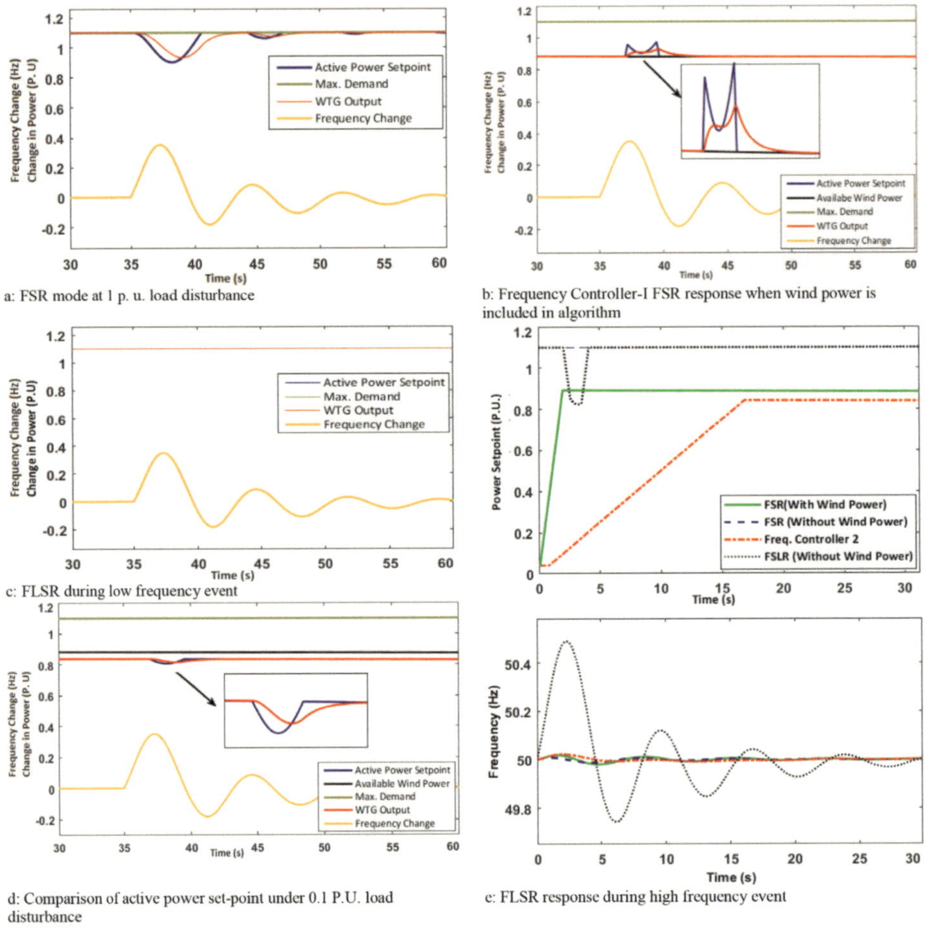

Figure 23.
Comparison of frequency controllers' response for step load change. (a) FSR mode at 1 p.u. Load disturbance. (b) Frequency controller-I FSR response when wind power is included in algorithm. (c) FLSR during low-frequency event. (d) Comparison of active power set point under 0.1 p.u. load disturbance. (e) FLSR response during high-frequency event.

is slightly increased under wind power and variable droop-based frequency controller-II. Frequency nadir point is nearly the same when type 3 VSWTG is working under frequency controller-I. Frequency controller-II performance in terms of frequency nadir point is slightly deteriorated under type 4 VSWTG integration during low load condition as shown in **Figure 25a**. Electrical power output from both type 3 and type 4 VSWTG is reduced during frequency deviations under low and high load disturbances as shown in **Figures 24c, d** and **25c, d**. Respective power set points provided by frequency controllers are shown in **Figure 24e, f**. During the frequency disturbance, when electrical power support is provided from VSWTG models, rotor speed remains above the minimum limit of 0.7 p.u. for both types of VSWTG model as shown in **Figures 24g** and **25g**.

The reduction in power output is highly dependent upon active power set point algorithm based on available wind power and variable droop. Wind power calculation through manufacturer provided curve applies an oversimplified approach where wind power is modeled primarily as the cube function of hub height wind speed alone, while practically other factors like wind shear and turbulence are also involved [40, 41]. Wind power forecasting involves conversion of atmospheric

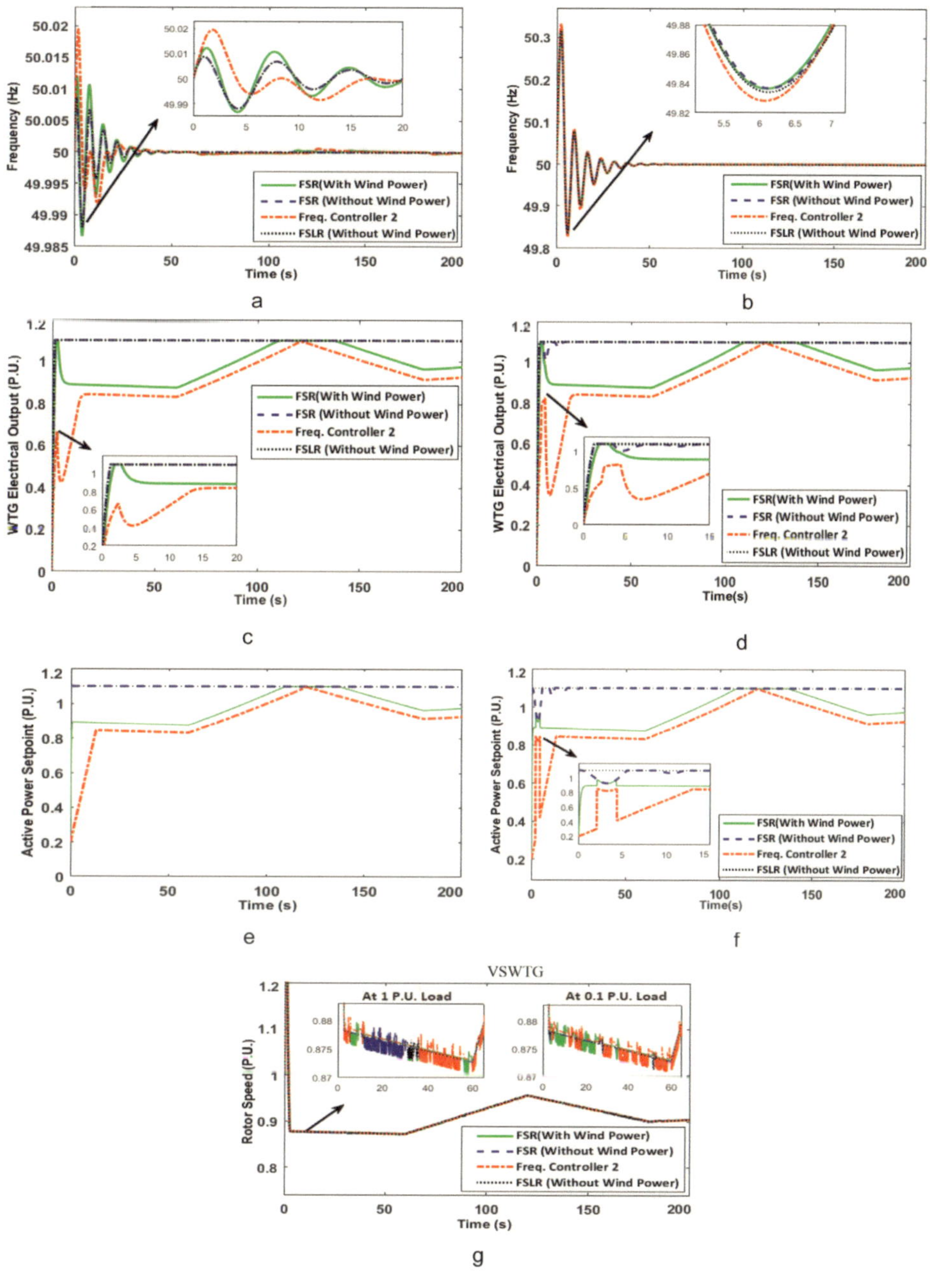

Figure 24.
Type 3 VSWTG frequency-sensitive response during step load change. (a) Type 3 VSWTG frequency response at 0.1 p.u. load disturbance. (b) Type 3 VSWTG frequency response at 1 p.u. load disturbance. (c) Type 3 VSWTG electrical power at 0.1 p.u. load disturbance. (d) Type 3 VSWTG electrical power at 1 p.u. load disturbance. (e) Active power set point at 0.1 p.u. load disturbance for type 3 VSWTG. (f) Active power set point at 1 p.u. load disturbance for type 3 VSWTG. (g) Type 3 VSWTG rotor speed.

forecasts into turbine power output forecasts. Inaccurate measurement and forecasting may highly affect wind turbine output and turbine life. 20% error in wind speed forecasting may introduce around 41% error in wind power output [42]. Anticipation of actual wind energy at time horizon less than 1 min is hard as wind power forecasting involves a lot of uncertainty due to spatial and temporal variability of wind fields and different forecasting tools. Accurate wind power forecasts,

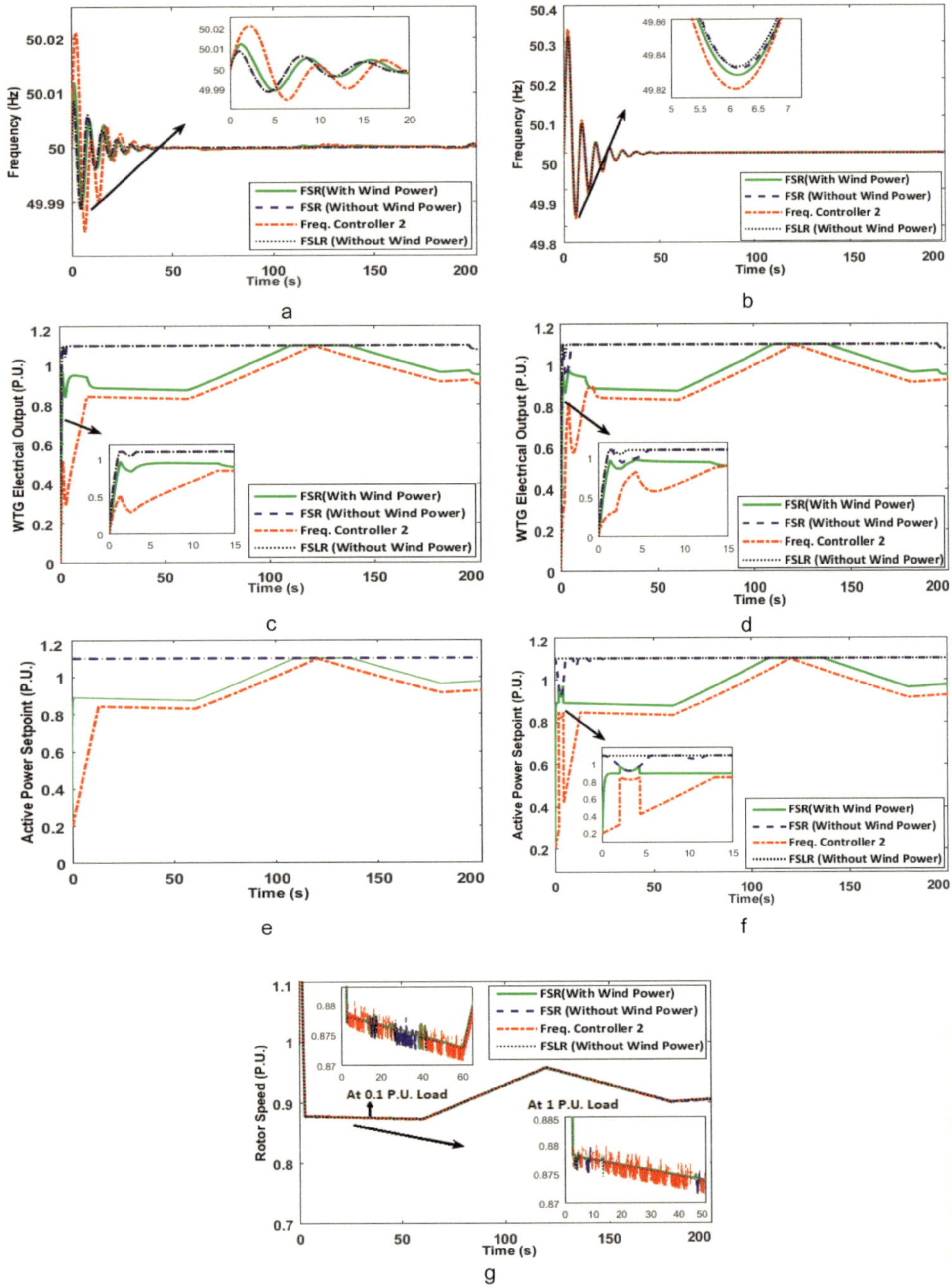

Figure 25.
Type 4 VSWTG frequency-sensitive response during step load change. (a) Type 4 VSWTG frequency response at 0.1 p.u. load disturbance. (b) Type 4 VSWTG frequency response at 1 p.u. load disturbance. (c) Type 4 VSWTG electrical power at 0.1 p.u. load disturbance. (d) Type 4 VSWTG electrical power at 1 p.u. load disturbance. (e) Active power set point at 0.1 p.u. load disturbance for type 4 VSWTG. (f) Active power set point at 1 p.u. load disturbance for type 4 VSWTG. (g) Type 4 VSWTG rotor speed.

related uncertainty, and their corresponding effect on wind turbine controller are computationally challenging and require a multiscale simulation approach [41]. Integration of wind forecasting will significantly increase the processing time of the turbine controller.

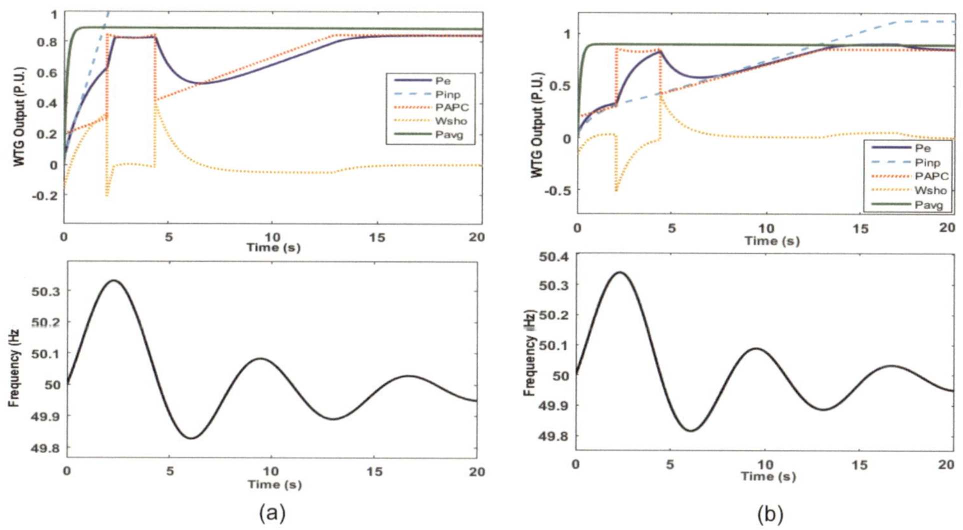

Figure 26.
Comparison of type 3 and type 4 VSWTG model internal power orders during frequency disturbance. (a) Type 3 VSWTG outputs at 1 p.u. load disturbance. (b) Type 4 VSWTG outputs at 1 p.u. load disturbance.

The generic model of type 3 and type 4 are different mainly in terms of electrical control model and different gain values for torque controller. Type 4 VSWTG generic model includes braking resistance which has no impact during normal frequency disturbance, while the braking resistance absorbs excessive energy when power order is larger than delivered energy to the grid. There is slight variation in final electrical power order for type 3 and type 4 VSWTG when wind power is included in algorithm and is shown in **Figure 26**. Final electrical power order is also dependent upon internal power set points and washout filters which are shown in **Figures 19** and **20**. The turbine control model sends a power order (P_{inp}) to the electrical control, requesting that the converter deliver this power to the grid. This power order is further altered by frequency-sensitive power set point (P_{APC}). The electrical control may or may not be successful in implementing this power order. **Figure 26a, b** shows the active power response of type 3 and type 4 VSWTG. Wind turbine final electrical order (P_e) initially follows the power order P_{inp} but then starts following the frequency controller power order $\mathrm{P_{APC}}$ as soon the frequency crosses the threshold. The washout filter power response rate limit (Wsho, orange line) transiently allows the power order variations from the (P_{APC}, red) through to the final power order (P_e). Due to the difference in torque controller gains and low-pass filter gain for type 3 and type 4 VSWTG, a comparatively faster matching response is observed for type 4 VSWTG control, where final electrical power order P_e jumps from minimum P_{inp} to maximum $\mathrm{P_{APC}}$ during frequency disturbance. Type 3 VSWTG control also closely follows the power order P_{APC} during frequency disturbance. Final power order P_e $=P_{APC}$ at around 15 s for type 3 VSWTG and around 18 s for type 4 VSWTG

3.3 Comparison of frequency controller-based type 3 VSWTG model with other frequency-sensitive models

Frequency droop model-I [43], droop model-II [44], and inertia droop model [18] are incorporated in basic torque control loop-based type 3 VSWTG model [16] which is shown as inset in **Figure 19** for comparison with modified control loop-based VSWG response when provided with frequency-sensitive power set point.

Figure 27 gives the frequency deviation obtained from single area controlled model with 0.1 per unit load disturbance. Due to frequency-responsive active power support from VSWTG, a clear improvement in area frequency deviation can be observed with proposed frequency-responsive VSWTG model integration. In single controlled area model, low-frequency deviation and low settling time of around 30–35 s are observed with frequency controller-based VSWTG integration. More overshoots and undershoots are observed for other droop-based VSWTG models and settling time of around 45–50 s. Superiority of grid code-compatible frequency response controllers is established through simulation results in terms of reduced settling time, improved ROCOF, and frequency nadir point.

Rate of change of frequency after a disturbance either due to load variation or generation imbalance is determined on the basis of system inertia and amount of imbalance and is given as

$$\frac{df}{dt} = \frac{\frac{-f_0}{2H_{conv}} * \Delta P}{S_{conv}} \tag{13}$$

The management of ROCOF is critical to maintaining power system frequency within the frequency operating standard and to maintaining the power system in a secure operating state. The amount of inertia required to maintain a particular ROCOF under different contingency is proportional to the contingency size. Lower inertia leads to a higher ROCOF. That means the frequency changes faster following a disturbance in a power system with less synchronous generation, and this could result in the loss of additional generation or load to arrest the frequency deviation when it occurs. There is no TSO's control over minimum system inertia, but ROCOF has to be within certain limits as per grid code. ROCOF obtained under different load disturbance scenario and 10% wind penetration is given in **Figure 27c**. Though the values obtained cannot be held indicative of real-time scenario which has interaction between components of varying electrical characteristics, a clear improvement is observed in ROCOF values for control area when proposed frequency-based controller-based Type 3 VSWTG is integrated with hydro power plant. A more detailed system will be required to analyze actual ROCOF changes.

Frequency nadir after any contingency is detrimental to primary frequency regulation for maintaining system stability. The primary frequency reserve adequacy criterion can be expressed as [45]

$$Freq_{Nadir} = (\Delta P_{Load}, PrimaryFrequencyReserve, F_{db}) \geq Freq_{min}, \tag{14}$$

where ΔP_{Load} is the maximum power loss during contingency, F_{db}(Hz) is the maximum governors' dead-band, $Freq_{Nadir}$ is the frequency nadir after the loss of ΔP_{Load}, and $Freq_{min}$ is the minimum frequency required. Wind power plant integration and its participation in frequency regulation services can be analyzed through frequency nadir points achieved under different combinations. Frequency nadir obtained for different frequency-responsive VSWTG models integrated in control area is shown in **Figure 27d**. Frequency nadir point as achieved under different load disturbance scenario indicates the superiority of grid code compatible frequency controllers over normal droop-based VSWTG model.

An important point to notice about other three droop-controlled models is in terms of wind plant power output during and after frequency response in LFC model. **Figure 27e, f** shows the zoomed version of electrical power output from VSWTG with different droop-controlled models during the initial frequency deviation in control area. Even though a variable wind speed is applied in simulation, a constant wind speed of 11.3 m/s is observed during approximately 20 s of frequency deviation.

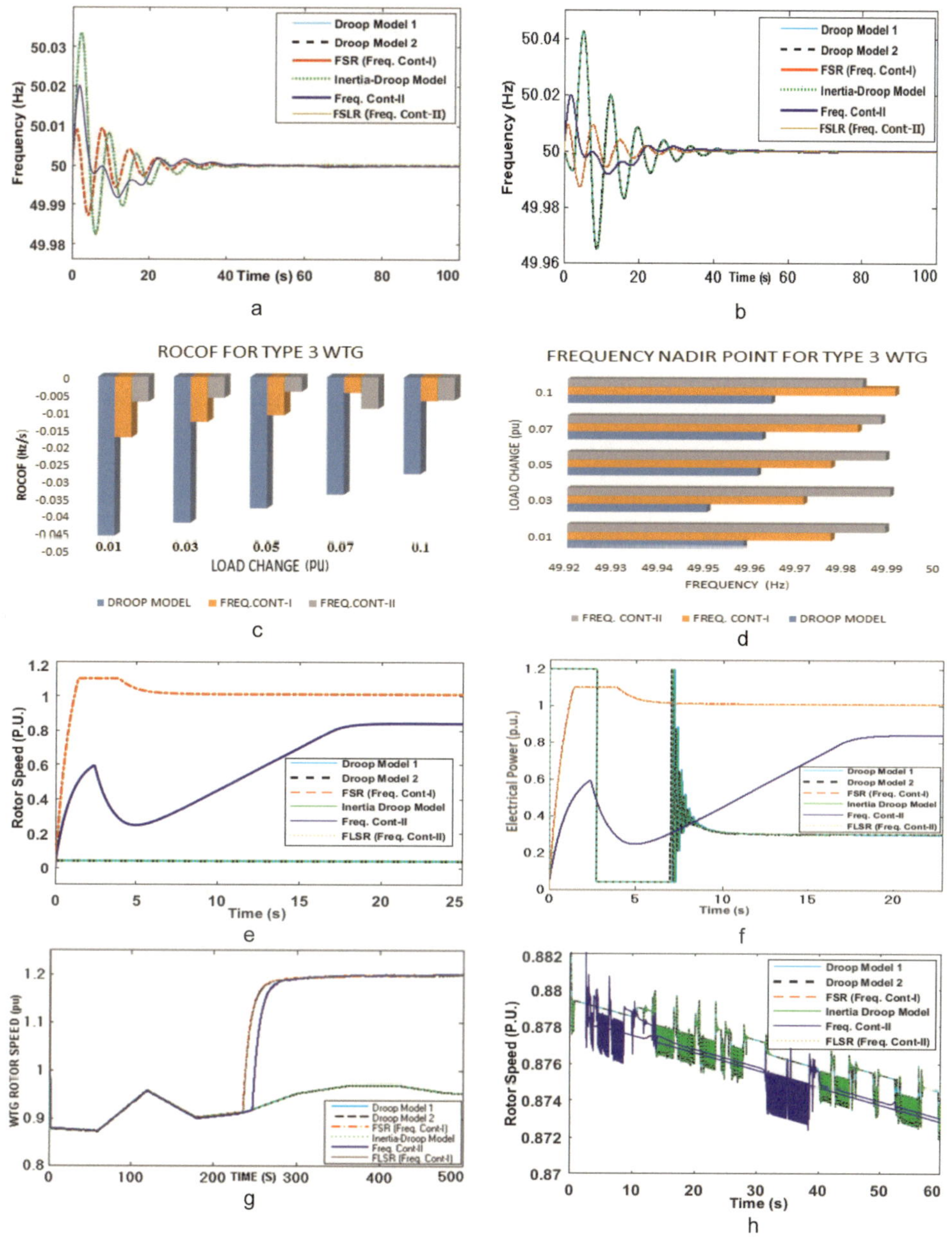

Figure 27.

Comparison of type 3 VSWTG modified model with other droop-based models incorporated with basic VSWTG model. (a) Frequency response of single area LFC control model for 0.1 p.u. Load disturbance, Req = 0.035 for all cases, torque controller gains (3, 0.6), no retuning applied for torque controller in simple VSWTG models. (b) Frequency response of single area LFC control model for 0.1 p.u. Load disturbance and Req = 0.035 in all cases, torque controller retuned for other droop-based VSWTG models. (c) Rate of change of frequency (ROCOF) comparison for different load change (p.u.) and 10% VSWTG penetration. (d) Frequency nadir comparison for different load change (p.u.) and 10% VSWTG penetration. (e) Electrical power from type 3 VSWTG with no retuning of torque controller for 0.1 p.u. Load disturbance. (f) Electrical power from type 3 VSWTG with retuned torque controller for 0.1 p.u. Load disturbance. (g) VSWTG rotor speed variations during frequency response. (h) VSWTG rotor speed variations during frequency response (Zoomed).

Considering a 0.1 p.u. load disturbance and VSWTG provided 0.05 p.u. active power support, control area sees this frequency support from a machine with 3.4 s inertia constant. This inertia constant will change to 7.17 s if 0.1 p.u. of extra support from VSWTG is assumed. VSWTG will replace other conventional generation (having equivalent inertia 6 s) if they provide an active power support of approximately 0.8 p.u. during frequency excursions. It can be observed that even with same load disturbance and same system inertia constant for all control areas, a different frequency response is obtained. All the three droop-based models attempt to provide increased electrical power up to 1.2 p.u. soon after detecting the frequency drop but fail to maintain net power output after frequency disturbance correction. Electrical power from basic VSWTG model finally settles at 0.4 p.u. as soon as frequency improves. This was also earlier confirmed in [16] and requires torque controller parameter retuning for increased electrical power after temporary frequency deviations.

We can notice a clear improvement in electrical power temporary support from VSWTG model incorporating frequency controller-I and controller-II. Following the active power set point generation algorithm, a constant active power set point equal to 1.01 p.u. is provided by frequency controller-I to ramp up the power during the moments of frequency deviations. Frequency controller-II also provides an increasing power reference set point of around 0.85 p.u. during the initial 20 s of frequency deviation and changes according to grid requirements and available power. VSWTG model with frequency controller-I is able to maintain electrical power around 1.1 p.u. during moments of initial frequency deviations, while VSWTG with frequency controller 2 has initial power drop up to 0.3 p.u. and then starts increasing and settles at around 1.1 p.u.

Figure 27g, h shows the change in rotor speed for Type 3 VSWTG model during frequency response in control area. All VSWTG models show reduced rotor speed maintained approximately 0.87 p.u. during those initial moments of frequency deviations till 20 s when extra electrical power from VSWTG is expected. Rotor speed of type 3 VSWTG with other droop models finally settles at around 0.95 p.u. with consequent reduced electrical output power to 0.4 p.u. Drop in rotor speed is observed due to imbalance in mechanical and electromechanical torque when electrical power is increasing for frequency controller-I and controller-II, but as soon as frequency deviation settles, an increase in rotor speed is also observed which settles at around 1.2 p.u. A minimum limit of 0.75 p.u. and maximum limit of 1.2 p.u. has been imposed in current studied VSWTG model.

4. Conclusions

There is an increasing demand for frequency control ancillary services due to high penetration of wind power plants in electrical network. This paper presented a modeling framework for frequency dependent active power set point generation in variable speed wind turbines and its corresponding effects on system frequency regulation response. Individual wind turbine can be made grid code frequency compatible by including additional active power set point generator output to the modified torque control loop of respective turbine. Active power set point generator applies designed power limitation on available wind power, rated turbine power, and TSO commanded power and provides set point to turbine. A grid frequency processor based on dynamic dead-band and moving averaged frequency filter is used to suppress noise frequency signals from passing to active power set point generator. Simulations were carried out for a single area control model with 10% wind penetration. Control area frequency responses as well as integrated VSWTG's electrical and rotor speed responses were compared with other frequency droop-based VSWTG model responses. Promising results in terms of improved settling time and better electrical power and rotor speed variations during frequency deviations were obtained for proposed frequency-responsive VSWTG model in comparison to other common droop-inertia-based model.

Grid processor presented in this study will be further improved and tested with low-pass IIR filter with a resettable strategy once measured frequency enters again. One per unit power limitation was applied in active power set point controller which will be further tested for more realistic power commands. Frequency response for area 1 was determined using classical linear hydraulic turbine model. It would be interesting to investigate the area frequency response with advanced inelastic nonlinear hydraulic turbine model interaction with frequency-responsive wind power plants.

A. Appendix

1. LFC system parameters used in single area LFC simulation

H_{system}	T_d	R
6	3.7814	0.0315
K_i (for single area model)	P_L (p.u.)	L_p (wind penetration)
−3.88	1–30%	10%

2. Droop model parameters

Inertia droop model: low-pass filter constant, 0.1 s; $K_{inertia}$, 2*H_{wind}; high-pass filter constant, 1 s; K_{droop}, 0.0315
Droop model-I: low-pass filter constant, 0.01 s; washout filter constant, 6 s; droop, 0.0315 Droop model 2: low-pass filter constant, 0.2 s; K_{droop}, 0.0315

3. Frequency grid processor parameters

Prefilter constant, 0.5 s; trend generator/filter constant, 8 s; upper dead-band, 0.015
Lower dead-band, 0.001; post filter constant, 0.5 s; droop, 0.0315

4. VSWTG parameters (differences are highlighted in bold)

Type 3 VSWTG (modified torque loop) with frequency response controller	Type 3 VSWTG (basic torque loop) with droop controllers
Aerodynamic model: $C_p(\lambda,\theta) = C_1\left(\frac{C_2}{\lambda_i} - C_3\theta - C_4\right)e^{\frac{-C5}{\lambda_i}} + C_6\lambda$ $\frac{1}{\lambda_i} = \frac{1}{\lambda+0.08\theta} - \frac{0.035}{\theta^3+1}$ c1 = 0.5176, c2 = 116, c3 = 0.4, c4 = 5, c5 = 21, and c6 = 0.0068 Pitch controller, K_{pp} = 150, K_{ip} = 25; pitch angle limitation, 0–27° Pitch compensation: K_{pc} = 3, K_{ic} = 30 **Inputs: P_{APC}, P_{inp}** Low-pass filter 2 time constant: 0.3 s **Torque controller: K_{ptrq} = 3, K_{itrq} = 0.6** Low-pass filter 4 time constant: 60s Power limitation: 0.04–1.1 p.u. **Washout filter constant: 1.0 s** Gen/converter model: Maximum current limitation: 1.1 p.u. Low-pass filter 3: 0.02 s	Aerodynamic model: -Same Pitch controller: same Angle limitation: same Pitch compensation, same; **Inputs, Pref = 1, P_{inp}** **Torque controller: K_{ptrq}, −271.5; K_{itrq}, −310.7** Low-pass filter (TC) constant: 5 s Power limitation: same **Washout filter not present** Gen/converter model: Maximum current limitation: 1.1 Low-pass filter: 0.02

Type 4 VSWTG has all the same parameters as that of Type 3 VSWTG Electrical control: E_{bst} (p.u.), 0.2; K_{dbr}, 10 Torque controller: $\mathbf{K_{ptrq} = 0.3}$, $\mathbf{K_{itrq} = 0.1}$ Low-pass filter time constant: 4 s	Type 4 basic VSWTG model has all the same parameters as that of modified loop-based type 4 VSWTG model except Pitch compensator taking **Pref = 1** as input Low-pass filter time constant: 0.05 s

Author details

Asma Aziz* and Aman Than Oo
Faculty of Science, Engineering and Built Environment, Deakin University, Australia

*Address all correspondence to: asma.aziz@deakin.edu.au

IntechOpen

© 2020 The Author(s). Licensee IntechOpen. Distributed under the terms of the Creative Commons Attribution - NonCommercial 4.0 License (https://creativecommons.org/licenses/by-nc/4.0/), which permits use, distribution and reproduction for non-commercial purposes, provided the original is properly cited. CC BY-NC

References

[1] AEMO. Power System Operations Event Report: Multiple Generator Disconnection and under Frequency Load Shedding. 2009. Available from: aemo.com.au

[2] Kundur P, Balu N, Lauby M. Power System Stability and Control. New York: McGraw-Hill; 1994

[3] Jonkman J, Butterfield S, Musial W, Scott G. Definition of a 5-MW Reference Wind Turbine for Offshore System Development, NREL/TP-500-38060. Golden, CO: National Renewable Energy Laboratory; 2009

[4] Pourbeik P, Ellis A, Sanchez-Gasca J, Kazachkov Y, Muljadi E, Senthil J, et al. Generic stability models for Type 3 & 4 wind turbine generators for WECC, In: Power and Energy Society General Meeting (PES). IEEE; 2013. p. 1, 5, 21, 22

[5] Sørensen B, Andersen B, Fortmann J, Johansen K, Pourbeik P. Overview, status, and outline of the new IEC 61400-27—Electrical simulation models for wind power generation. In: International Workshop on Large-Scale Integration of Wind Power into Power Systems as Well as on Transmission Networks for Offshore Wind Power Farms; 2011

[6] WECC. WECC Wind Power Plant Dynamic Modeling Guide, WECC Renewable Energy Modeling Task Force. 2010. Available from: renew-ne.org

[7] National Grid. Mandatory Frequency Response | National Grid [online]. 2016. Available from: http://www2.nationalgrid.com

[8] AEMC. Feasibility of Fast Frequency Response Obligations of New Generators. 2017. Available from: http://www.aemc.gov.au/getattachment/661d5402-3ce5-4775-bb8a-9965f6d93a94/AECOM-report-Feasibility-of-FFR-obligations-of-new.aspx

[9] de Almeida R, Lopes JP. Participation of doubly fed induction wind generators in system frequency regulation. IEEE Transactions on Power Systems. 2007;**22**:944-950

[10] Juankorena X, Esandi I, Lopez J, Marroyo L. Method to enable variable speed wind turbine primary regulation. In: International Conference on Power Engineering, Energy and Electrical Drives; 2009. pp. 495-500

[11] Liserre M, C'ardenas R, Molinas M, Rodriguez J. Overview of multi-MW wind turbines and wind parks. IEEE Transactions on Industrial Electronics. 2011;**58**(4):1081-1095

[12] Morren J, de Haan S, Kling W, Ferreira J. Wind turbines emulating inertia and supporting primary frequency control. IEEE Transactions on Power Systems. 2006;**21**(1):433-434

[13] Gowaid I, El-Zawawi A, El-Gammal M. Improved inertia and frequency support from grid-connected DFIG wind farms. In: Power Systems Conference and Exposition (PSCE). IEEE/PES; 2011. pp. 1-9

[14] Ma H, Chowdhury B. Working towards frequency regulation with wind plants: Combined control approaches. IET Renewable Power Generation. 2010;**4**(4):308-316

[15] Hansen AD. Evaluation of Power control with different electrical and control concepts of wind farms. Technical Report. Roskilde, Denmark: Project UpWind; 2010

[16] Ullah NR, Thiringer T, Karlsson D. Temporary primary frequency control support by variable speed wind turbines

—Potential and applications. IEEE Transactions on Power Systems. 2008; **23**(2):601-612

[17] Tarnowski GC, Kjar PC, Sorensen PE, Ostergaard J. Variable speed wind turbines capability for temporary over-production. In: IEEE Power & Energy Society General Meeting; Calgary, AB; 2009. pp. 1-7

[18] Ramtharan G, Ekanayake JB, Jenkins N. Frequency support from doubly fed induction generator wind turbines. IET Renewable Power Generation. 2007;**1**:3-9

[19] Attya AB, Hartkopf T. Wind farms dispatching to manage the activation of frequency support algorithms embedded in connected wind turbines. International Journal of Electrical Power & Energy Systems. 2013;**53**:923-936

[20] Hughes FM, Anaya-Lara O, Jenkins N, Strbac G. Control of DFIG-based wind generation for power network support. IEEE Transactions on Power Systems. 2005;**20**:1958-1966

[21] Mauricio JM, Marano A, Gomez-Exposito A, Ramos JLM. Frequency regulation contribution through variable speed wind energy conversion systems. IEEE Transactions on Power Systems. 2009;**24**:173-180

[22] Gautam D, Goel L, Ayyanar R, Vittal V, Harbour T. Control strategy to mitigate the impact of reduced inertia due to doubly fed induction generators on large power systems. IEEE Transactions on Power Systems. 2011; **26**:214-224

[23] Aziz A, Shafiullah GM, Stojcevski A, MTO A. Participation of DFIG based wind energy system in load frequency control of interconnected multi generation power system. In: Power Engineering Conference (AUPEC). Perth, WA: Australasian Universities, 2014. pp. 1-6

[24] Conroy JF, Watson R. Frequency response capability of full converter wind turbine generators in comparison to conventional generation. IEEE Transactions on Power Systems. 2008; **23**(2):649-656

[25] Grunnet JD, Soltani M, Knudsen T, Kragelund MN, Bak T. Aeolus toolbox for dynamics wind farm model, simulation and control. In: European Wind Energy Conference and Exhibition, EWEC 2010: Conference Proceedings; 2010

[26] Aho J, Buckspan A, Laks J, Fleming P, Jeong Y, Dunne F, et al. A tutorial of wind turbine control for supporting grid frequency through active power control. In: Proceedings of the 2012 American Control Conference, Montreal; 2012

[27] Janssens NA, Lambin G, Bragard N. Active power control strategies of DFIG wind turbines. In: 2007 IEEE Power Tech, Lausanne; 2007. pp. 516-521

[28] Tarnowski GC, Kjær PC, Dalsgaard S, Nyborg A. Regulation and frequency response service capability of modern wind power plants. In: Proceedings of the Power and Energy Society General Meeting, 25–29 July 2010. Minneapolis, USA: IEEE; 2010

[29] Hepner S, Wihler A. Method for controlling an output of an electrical power plant. US6118187 A; 2000

[30] Smith S. The Scientist and engineer's Guide to Digital Signal Processing. San Diego, CA: California Technical Publications; 1997

[31] ERA, 2014 ancillary service standards and requirements study, EY report to the Independent Market Operator, 2014. Available from: www.erawa.com.au/cproot/14770/2/EY%20Final%20Report.pdf

[32] EY. Ancillary Service Standards and Requirements Study, Perth. 2014. Available from: wa.aemo.com.au

[33] North American Electric Reliability Council. NERC OPERATING MANUAL. NERC; 2004. Available at: http://quantlabs.net/academy/download/free_quant_instituitional_books_/[NERC]%20NERC%20Operating%20Manual%20-%20June%202004.pdf

[34] Gibbard M, Vowles D. Simplified 14-Generator Model of the SE Australian Power System. South Australia: The University of Adelaide; 2008

[35] The Grid Code. Issue 4, Revision 10. nationalgridESO. GC0110: LFSM-O compliance requirements for Type As and B PGMs, Stage 02: Workgroup Report. 2019. UK: National Grid Electricity Transmission PLC; 2012. Available at: www.nationalgrideso.com/document/117576/download

[36] Manwell J, McGowan J, Rogers A. Wind Energy Explained: Theory, Design and Application. Chichester: Wiley; 2011

[37] Clark K, Miller NW, Sanchez-Gasca JJ. Modelling of GE Wind Turbine-Generators for Grid Studies, Version 4.5. General Electric International, Inc; 2010

[38] Aziz A, Maung Than Oo A, Stojcevski A. Full converter based wind turbine generator system generic modelling: Variations and applicability. Sustainable Energy Technologies and Assessments. 2016;**14**:46-62

[39] Chan ML, Dunlop RD, Schweppe F. Dynamic equivalents for average system frequency behaviour following major disturbances. IEEE Transactions on Power Apparatus and Systems. 1972; **PAS-91**(4):1637-1642

[40] Miller W, Heller A. Breath of Air. In: Meteorological Technology International Journal. U.K. 2015. p. 22. Available from: http://www.mbw.ch/wp-content/uploads/2015/05

[41] Str.llnl.gov. Predicting Wind Power with Greater Accuracy [Online]. 2017. Available from: https://str.llnl.gov/april-2014/miller

[42] Sharman H. Why wind power works for Denmark. Proceedings of the Institution of Civil Engineers: Civil Engineering. 2015;**158**(2):66-72

[43] Mauricio JM, Marano A, Gómez-Expósito A, Martínez Ramos JL. Frequency regulation contribution through variable-speed wind energy conversion systems. IEEE Transactions on Power Systems. Feb. 2009;**24**(1): 173-180

[44] Yingcheng X, Nengling T. System frequency regulation in doubly fed induction generators. International Journal of Electrical Power & Energy Systems. 2012;**43**(1):977-983

[45] Chavez H, Baldick R, Matevosyan J. The joint adequacy of AGC and primary frequency response in single balancing authority systems. IEEE Transactions on Sustainable Energy. 2015;**6**(3):959-966

Chapter 8

Nonlinear and Sampled Data Control of Wind Turbine

Marwa Hassan

Abstract

This chapter aims to investigate the effectiveness of the nonlinear control-based model and the sampled-data design through the power system application. In particular, the study will focus on a model of a wind turbine system fed by a doubly fed induction generator (DFIG). First of all, a detail dynamical model of a DFIG-based wind-turbine grid-connected system is presented in the direct and quadratic synchronous reference frame. Afterward, mathematical modeling is performed for the nonlinear and sampled data systems. The nonlinear control will ensure the reproduction of the rotor direct and quadratic current that converge to the reference signal generated from. The proposed sampled-data system is built upon the nonlinear model and is introduced as an alternative of the classical discrete control which is known as emulation design. The simulation's results will show that implementing the approximate feedback will yield better results than the one obtained from the mere emulation.

Keywords: sampled data model, nonlinear control, renewable energy, wind turbine, feedback

1. Introduction

In recent years global warming emissions have been one of the most important topics discussed by the researcher. The carbon dioxide causes harmful impacts on the environment as it acts like a blanket that traps heat. According to the latest survey done in the United States, about 29% of global warming emissions are caused by fossil fuel used in the electricity sector [1–3]. Therefore it becomes more essential to look for alternative sources. Renewable energy sources especially wind energy produce little to no global warming emissions as burning natural gas for electricity releases between 0.6 and 2 pounds of carbon dioxide equivalent per kilowatt-hour (CO_2E/kWh); coal emits between 1.4 and 3.6 pounds of CO_2E/kWh. Wind, on the other hand, is responsible for only 0.02–0.04 pounds of CO_2E/kWh on a life-cycle basis [3]. In conclusion, Wind energy represents one of the fastest-growing energy sources in the world due to it is a major advantage in terms of cost and effectiveness [4–8]. The wind farm system based on the doubly fed induction generator (DFIG) will be studied in this chapter. In simple words, the DFIG is a generator that has its rotor winding connected to the grid via slip rings and back-to-back voltage source converters that control both the rotor and the grid currents. Thus, rotor frequency can freely differ from the grid frequency. DFIG has become more effective in the industry in the last few years due to its advantage compared to the permanent magnet synchronous generator as it provides better results when compared to cost, size, and weight. A lot of researches have been carried out in the area of modeling and controlling of stator/rotor flux in DFIG [9–11]. Most of these researches face some drawbacks due to the nonlinear nature of the DFIG. Examples can be found on [12–19]. In this chapter, we aim to introduce a nonlinear control technique that aims to reproduce an output signal that converges to a distinct reference signal. Later on, the sampled data control technique will be applied as an alternative solution of the emulation based technique that is usually applied to the nonlinear system. A few examples of the sampled data are illustrated in [20–22]. The chapter is structured as follow: Section 2 recalls the modeling considerations of the doubly fed induction generator while Section 3 illustrates the nonlinear control approach. Section 4 presents the Grid side command model. Finally, Section 5 presents the sampled data model and the obtained results while Section 6 concludes the paper and formulates further research directions.

2. Modeling

The modeling of the doubly fed induction generator (DFIG) will be recalled in this section. These equations will be used to design the nonlinear control system.

2.1 Doubly fed induction generator model

1. Stator equations

$$V_{sd} = R_s i_{sd} + \frac{d}{dt}\lambda_{sd} - \lambda_{sq} W_s \tag{1}$$

$$V_{sq} = R_s i_{sq} + \frac{d}{dt}\lambda_{sq} + \lambda_{sd} W_s \tag{2}$$

$$\lambda_{sd} = L_s i_{sd} + M i_{rd} \tag{3}$$

$$\lambda_{sq} = L_s i_{sq} + M i_{rq} \tag{4}$$

2. Rotor equations

$$V_{rd} = R_r i_{rd} + \frac{d}{dt}\lambda_{rd} - \phi_{rq} W_r \tag{5}$$

$$V_{rq} = R_r i_{rq} + \frac{d}{dt}\lambda_{rq} + \phi_{rd} W_r \tag{6}$$

$$\lambda_{rd} = L_r i_{rd} + M i_{sd} \tag{7}$$

$$\lambda_{rq} = L_r i_{rq} + M i_{sq} \tag{8}$$

$$W_r = g.W_s \tag{9}$$

with

$$i_{sd} = \frac{\lambda_{sd} - M i_r d}{L_s} \tag{10}$$

$$i_{sq} = \frac{-M i_{rq}}{L_s} \tag{11}$$

where R_s and R_r are, respectively, the stator and rotor phase resistances, Ls, Lr, M Stator and rotor per phase winding and magnetizing inductances and W_s, W_r are the stator and rotor speed pair pole number. The direct and quadrate stator and rotor currents are respectively represented as i_{sd}, i_{sq}, i_{rd} and i_{rq}. The electromagnetic torque is presented by the following equation:

$$J\frac{dW_r}{dt}+f_r W_r = c_{em} - c_r \tag{12}$$

$$C_{em} = p\left(\lambda_{rq} i_{rd} - \lambda_{rd} i_{rq}\right) \tag{13}$$

The system now will be modeled with respect to the rotor side direct and quadratic (d, q) synchronous reference frame. The input in such case are I_{rd} and I_{rq}.

First the system expression w.r.t d axis frame:

$$v_{rd} = R_r i_{rd} + \frac{d}{dt}(L_r i_{rd} + M i_{sd}) - (L_r i_{rq} + M i_{sq}) W_r \tag{14}$$

$$= R_r i_{rd} + \frac{d}{dt} i_{rd}\left(L_r - \frac{M^2}{L_s}\right) - L_r i_{rq} W_r - M W_r\left(\frac{-M i_{rq}}{L_s}\right) \tag{15}$$

$$R_r i_{rd} + L_r \frac{d}{dt} i_{rd}\left(1 - \frac{M^2}{L_s L_r}\right) - L_r i_{rq} W_r + \frac{M^2}{L_s} W_r i_{rq} \tag{16}$$

$$= R_r i_{rd} + L_r \dot{i}_{rd}\left(1 - \frac{M^2}{L_s L_r}\right) - L_r W_r\left(1 - \frac{M^2}{L_s L_r}\right) i_{rq} \tag{17}$$

$$= R_r i_{rd} + L_r \Lambda \dot{i}_{rd} - L_r \Lambda W_r i_{rq} \tag{18}$$

$$\dot{i}_{rd} = \frac{1}{L_r \Lambda} v_{rd} - \frac{R_r}{L_r \Lambda} i_{rd} + w_r i_{rq} \tag{19}$$

$$\dot{i}_{rd} = \frac{1}{L_r \Lambda} v_{rd} - \frac{1}{\mathscr{T}\Lambda} i_{rd} + w_r i_{rq} \tag{20}$$

with $\Lambda = \left(1 - \frac{M^2}{L_s L_r}\right) \mathscr{T} = \frac{R_r}{L_r}$.

Now consider q axis frame:

$$V_{rq} = R_r i_{rq} + \frac{d}{dt} \lambda_{rq} + \lambda_{rd} W_r \tag{21}$$

$$= R_r i_{rq} + \frac{d}{dt}\left(L_r i_{rq} - \frac{M^2}{L_s} i_{rq}\right) + W_r\left(L_r i_{rd} - \frac{M^2 i_{rd}}{L_S}\right) \tag{22}$$

$$= R_r i_{rq} + L_r \dot{i}_{rq}\left(1 - \frac{M^2}{L_s L_r} + L_r W_r\left(1 - \frac{M^2}{L_s L_r}\right) i_{rd}\right. \tag{23}$$

$$= R_r i_{rq} + L_r \Lambda \dot{i}_{rq} - L_r \Lambda W_r i_{rd} \tag{24}$$

$$\dot{i}_{rq} = \frac{1}{L_r \Lambda} v_{rq} - \frac{1}{\mathscr{T}\Lambda} i_{rq} - w_r i_{rd} \tag{25}$$

Finally we obtain the speed from the torque equation as:

$$\dot{W}_r = -\frac{f_r}{J}W_r + \frac{p}{J}\lambda_{rq}i_{rd} - \frac{p}{J}\lambda_{rd}i_{rq} \tag{26}$$

3. Nonlinear control of the DFIG

First, we start by putting the model in the standard nonlinear form. Recalling from modeling, the system is introduced in the condensed nonlinear form:

$$\Sigma_C : \begin{cases} \dot{x} = f(x) + g_1(x)u_1 + g_2(x)u_2, & x \in \mathbb{R}^n, u \in \mathbb{R}^n \\ y = h(x). \end{cases} \tag{27}$$

where, $X = [x_1 \quad x_2 \quad x_3]^T = \left[i_{rd} \quad i_{rq} \quad W_r\right]^T$, $U = [u_1 \quad u_2]^T = \left[v_{rd} \quad v_{rq}\right]^T$. The function $f(x)$, $g(x)$ are smooth vector fields and the output function $h(x)$ is a smooth scalar function.

$$f(x) = \begin{pmatrix} -\frac{1}{\mathcal{T}\Lambda}x_1 + x_2x_3 \\ -\frac{1}{\mathcal{T}\Lambda}x_2 - x_2x_3 \\ -\frac{f_r}{J}x_3 + \frac{p}{J}\phi_{rq}x_1 - \frac{p}{J}\phi_{rd}x_2 \end{pmatrix} \tag{28}$$

$$g_1(x) = \begin{pmatrix} \frac{1}{\mathcal{T}\Lambda} \\ 0 \\ 0 \end{pmatrix}, \quad g_2(x) = \begin{pmatrix} 0 \\ \frac{1}{\mathcal{T}\Lambda} \\ 0 \end{pmatrix}. \tag{29}$$

Since the purpose of this study is to control the rotor side converter current, the output was chosen as $h(x) = \left[i_{rd}, i_{rq}\right]^T$.

Remark 1. *According to the previous results obtained by Isidori, A multi variable nonlinear system in the form of (36) has a relative degree $r_1, \ldots, r_m$ at point x^0 if $L_{g_j}L_f^k h_i(x) = 0$ for all $1 \leqslant j \leqslant m$, for all $1 \leqslant i \leqslant m$, for all $k \leqslant r_i - 1$, and for all neighbor of x^0.*

Following the same definition, it can be easily verified that the system relative degree w.r.t the outputs $r = 2$.

3.1 Control of d-axis rotor current

In order to track rotor current i_rq we assume that the system is only affected by u_1 and $u_2 = 0$.

$$\dot{x} = f(x) + g_1(x) \tag{30}$$

$$y = h_1(x) = i_{rd} \tag{31}$$

The system relative degree w.r.t the output $r = 1$. Now we apply a coordinate transformation and introduce the system in to the normal form.

$$\phi(x) = \begin{pmatrix} z_1 = x_1 \\ \eta_1 = x_2 \\ \eta_2 = x_3, L_g.\eta = 0. \end{pmatrix} \tag{32}$$

$$\begin{cases} \dot{z}_1 = -\frac{1}{\mathscr{T}\Lambda} z_1 + \eta_1\eta_2 + \frac{1}{\mathscr{T}\Lambda} u_1 \\ \dot{\eta}_1 = -\frac{1}{\mathscr{T}\Lambda}\eta_1 - z_1\eta_2 \\ \dot{\eta}_2 = p\phi_{rd} z_1 - p\phi_{rd}\eta_1 - \frac{f_r}{J} eta_2 \end{cases} \tag{33}$$

After applying the proper control law in the form of $u = \mathscr{T}\Lambda\left(-\frac{1}{\mathscr{T}\Lambda}x_1 + x_2x_3 + x_1^r - c_0x_1\right)$ where x_1^r, c_0 represents the rotor current desired value and the chosen zero, we obtain the desired output.

3.2 Control of q-axis rotor current

In this case the effect of u_2 is studied

$$f(x) = \begin{pmatrix} -\frac{1}{\mathscr{T}\Lambda} x_1 + x_2 x_3 \\ -\frac{1}{\mathscr{T}\Lambda} x_2 - x_2 x_3 \\ -\frac{f_r}{J} x_3 + \frac{p}{J}\phi_{rq} x_1 - \frac{p}{J}\phi_{rd} x_2 \end{pmatrix} g_2(x) = \begin{pmatrix} 0 \\ \frac{1}{\mathscr{T}\Lambda} \\ 0 \end{pmatrix} \tag{34}$$

$$y = h_2(x) = i_{rq}. \tag{35}$$

The system relative degree $r_q = 1$. The coordinate transformation and the normal take the form of

$$\phi(x) = \begin{pmatrix} z_1 = x_2 \\ \eta_1 = x_3 \\ \eta_2 = x_1 \end{pmatrix} \tag{36}$$

$$\begin{cases} \dot{z}_1 = -\frac{1}{\mathscr{T}\Lambda} z_1 + \eta_1\eta_2 + \frac{1}{\mathscr{T}\Lambda} u_2 \\ \dot{\eta}_1 = p\phi_{rd}\eta_1 - p\phi_{rd} z_1 - \frac{f_r}{J}\eta_1 \\ \dot{\eta}_2 = -\frac{1}{\mathscr{T}\Lambda}\eta_2 - z_1\eta_2. \end{cases} \tag{37}$$

The input

$$u = \mathscr{T}\Lambda\left(-\frac{1}{\mathscr{T}\Lambda}x_2 - x_1x_3 + x_2^r - c_0x_2\right). \tag{38}$$

3.3 Simulation and results

This section presents the evaluation of the performance of the proposed technique. Two cases were developed. The first case study based on the Doubly Fed Induction Generator model while the second one studies the gird side converter command model when the power factor is set to unity. **Table 1** presents the parameters of the DIFG parameters. The Bitz limit at which the maximum efficiency is obtained for the first case study is shown in **Figure 1**. In this case, the optimal point corresponds to Beta angle is equal to zero. Two feedbacks were

The DFIG data of a typical 3.6 MW generator	
Power	7 kW
Efficiency at rated speed	79%
Voltage	690 V
Locked rotor voltage	1000 V
Operation speed range	2000 rpm
Power factor	0.90 cap
Rotor Resistance	1 Ω
Rotor Inductance	0.2 mH
Stator Resistance	0.5 Ω
Stator Inductance	0.001 mH
Mutual inductance	0.078 H
Number of poles	4
Inertia moment	0.3125 Nms^2

Table 1.
The DFIG data sheet.

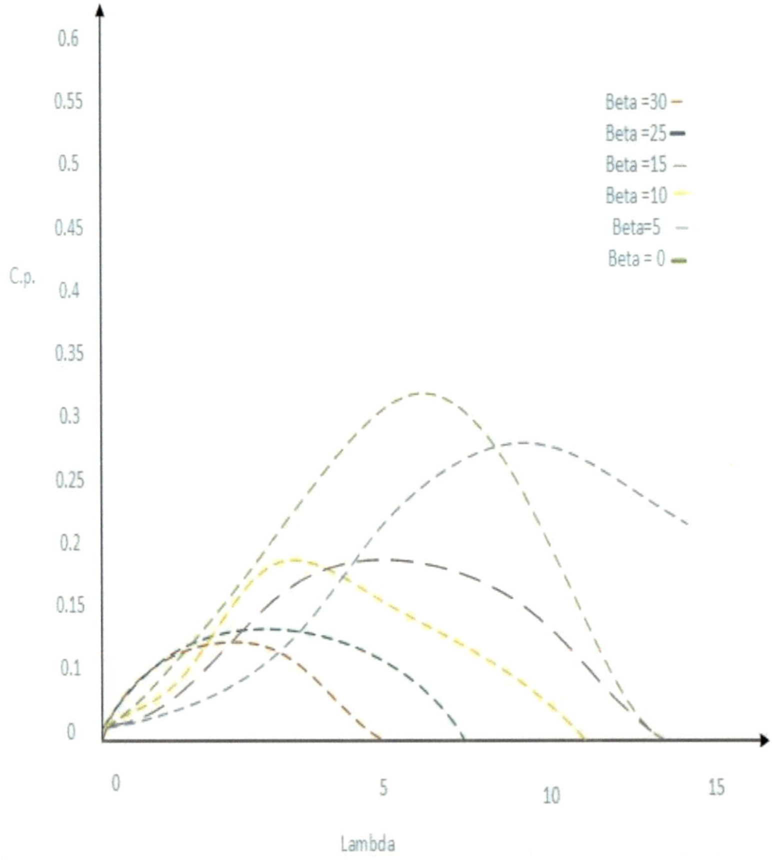

Figure 1.
Power coefficient curve.

applied in this stage in the sake of evaluating proposed control strategy. The primary feedback was applied in the direct axis frame with an input value

$$u = \mathcal{T}\Lambda\left(-\frac{1}{\mathcal{T}\Lambda}x_1 + x_2x_3 + x_1^r - 1000x_1\right) \tag{39}$$

Figure 2 shows the result of the rotor side reference signal and the generated current signals after applying the feedback. It can be noticed that the proposed control technique succeeded in reproducing a current signal that coincides with the required reference signal. As for the quadratic axis frame another feedback was designed to track the required current signal.

$$u = \mathcal{T}\Lambda\left(-\frac{1}{\mathcal{T}\Lambda}x_2 - x_1x_3 + x_2^r - 1200x_2\right) \tag{40}$$

The applied input will produce a signal that follows the reference signal (see **Figure 3**). **Figure 4** presents the continuous bus voltage of the DFIG regulated to the standard reference voltage fixed at 1000 V. It is clear that in spite of fluctuation of the wind the voltage remains stationary which is considered a major advantage as the system will be affected by the harmonics.

In the next half of this chapter, the sampled data design techniques and its application will be discussed. For the simplicity of the design, we choose to set the power factor to unity such that the system is converted to a SISO system. In such a case the Grid Side Converter command model is studied.

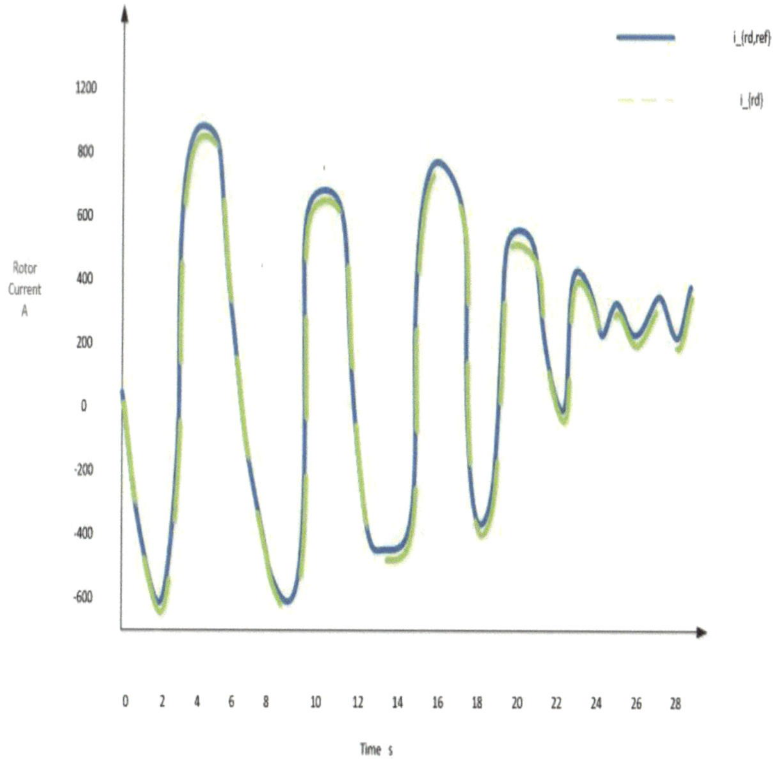

Figure 2.
Doubly fed induction generator i_{rd} rotor current.

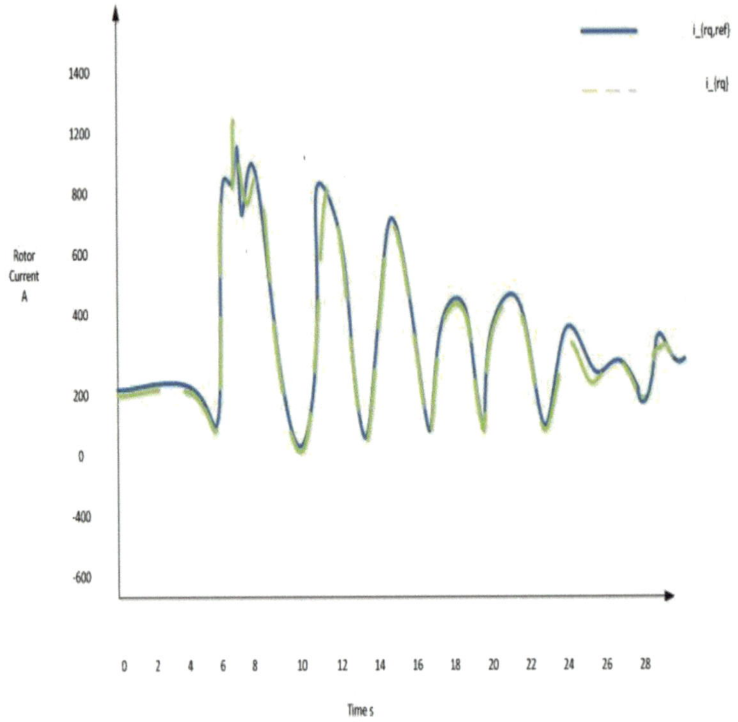

Figure 3.
Doubly fed induction generator i_{rq} rotor current.

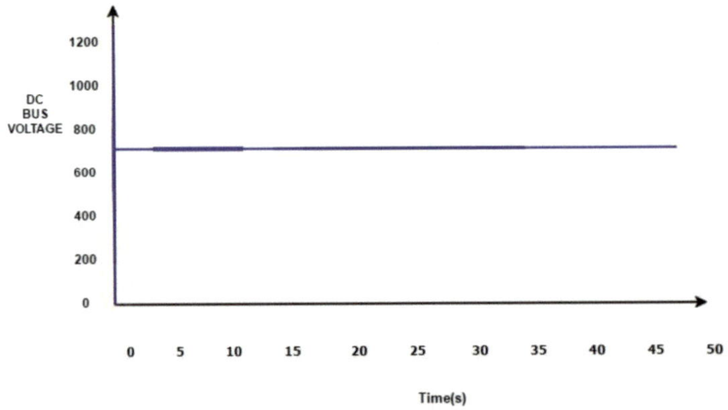

Figure 4.
Doubly fed induction generator continuous bus voltage.

4. Grid side converter command model

$$v_{fd} = R_f i_{fd} + L_f \frac{d}{dt} i_{fd} - W_r L_f i_{fq} - V_{Gd} \tag{41}$$

$$v_{fq} = R_f i_{fq} + L_f \frac{d}{dt} i_{fq} + W_r L_f i_{fd} - V_{Gq} \tag{42}$$

$$\dot{i}_{fd} = \frac{1}{L_f} \left(-R_f i_{fd} + v_{fd} + W_r L_f i_{fq} + V_{Gd} \right) \tag{43}$$

$$\dot{i}_{fq} = \frac{1}{L_f}\left(-R_f i_{fq} + v_{fq} - W_r L_f i_{fd} + V_{Gq}\right) \tag{44}$$

with V_{Gd}, V_{Gq} indicated the input voltage of the AC-DC converter in the direct and quadrature frame. The electric network components of voltage and current on the AC side for both the direct and quadrature frame are given by v_{fd}, v_{fq}, i_{fd} and i_{fq} respectively, while the L_f referred to the inductance of the system. The active and reactive power is expressed as:

$$P = V_{Gd} i_{fd} + V_{Gq} i_{fq} \tag{45}$$

$$Q = V_{Gq} i_{fq} - V_{Gd} i_{fd}. \tag{46}$$

Remark 2. *Through setting the power factor to be* 1 *and neglecting the filter losses one can get the following expression* $V_{Gd} = V_{fd} = V_G, V_{Gq} = V_{fq} = 0$, *leading the active and reactive power to be* $P_f = V_G i_{fd}$ *and* $Q_f = -V_G i_{fq}$.

5. Nonlinear grid side converter model

Referring to Remark 2 we know that through setting the power factor to unity we get $V_{Gd} = V_{fd} = V_G, V_{Gq} = V_{fq} = 0$. In such a case the system is converted into single input-single output system in the form:

$$f(x) = \begin{pmatrix} -\frac{R_f}{L_f} x_1 + x_3 x_2 \\ -\frac{R_f}{L_f} x_2 - x_3 x_1 \\ -\frac{f_r}{J} x_3 - \frac{p}{J} \phi_{rd} x_2 \end{pmatrix}, \quad g(x) = \begin{pmatrix} \frac{2}{L_f} \\ 0 \\ 0 \end{pmatrix} \tag{47}$$

$$y = x_1. \tag{48}$$

where x: state vector = $\begin{bmatrix} i_{fd} & i_{fq} & W_r \end{bmatrix}^T \quad U = \begin{bmatrix} u_1 & u_2 \end{bmatrix}^T$.

5.1 Nonlinear modeling and control of the quadratic axis control

In this part, the asymptotic output tracking technique will be studied in the quadratic axis frame. The system has a well define the relative degree of $r = 1$. Consequently one can apply a coordinate transformation in the form $\Gamma(x) = \begin{pmatrix} x_1 \\ x_3 \\ x_2 \end{pmatrix}$.

The state space description in the new coordinates

$$\begin{cases} \dot{z}_1 = a(z,\eta) + b(z,\eta) \\ \dot{\eta}_1 = \frac{f_r}{J}\eta_1 - \frac{p}{J}\phi_{rd}\eta_2 \\ \dot{\eta}_2 = -\frac{R_f}{L_f}\eta_1 - \eta_2 z_1. \end{cases} \tag{49}$$

Remark 3. *The system has a stable zero dynamics. In fact by calculating the Jacobian matrix Q which describes the linear approximation at $\eta = 0$ of the zero dynamics of the original nonlinear system*

$$Q = \begin{pmatrix} \frac{f_r}{J} & -\frac{p}{J}\phi_{rd} \\ -\frac{R_f}{L_f}b & 0 \end{pmatrix} \tag{50}$$

we can see that the matrix is nonsingular. Hence the zero dynamics are asymptotically stable. The stability of the zero dynamics will depend on the parameters of the DFIG.

The stator of the DFIG was directly connected to the grid while its rotor was connected to it via a cascade (Rectifier, Inverter, and Filter). To evaluate the grid side model the power factor was set to one, thus only the quadratic rotor current will be produced. The voltage on the output of the inverter will suffer from disturbance signals formed by the original frequency $f = 50$ Hz and other signals. A passive R-L filter was used to eliminate harmonics. The input in the form $u = \frac{1}{a(z,\eta)}(-b(z,\eta) + c_0 z_1)$ ensures the reproduction of an output i_{rq} that will track the required reference signal. **Figure 5** depicts that the system nonlinear controller has reproduced an output that will converge asymptotically to the required reference signals and minimizes the effect of disturbance.

5.2 Feedback design under sampling

We now address the problem of preserving under system behavior under sampling. In fact, considering $u(t) \in U_T$ and $y(t) = y(kT)$ for $t \in [kT, (k+1)T]$ (T the sampling period). Now we compute the single-rate sampled data equivalent model of (43)

$$x_{k+1} = F^T(x_k, u_k) \tag{51}$$

$$y_k = h(x_k) \tag{52}$$

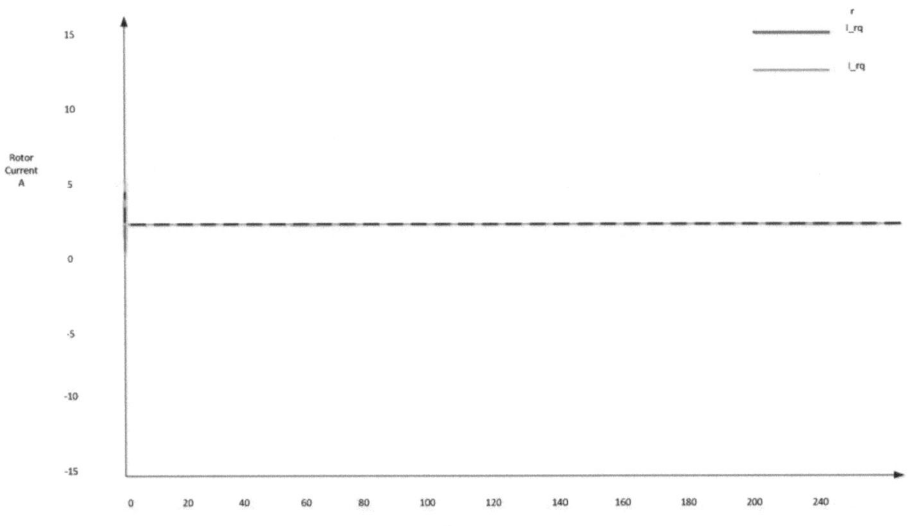

Figure 5.
Nonlinear control applied to rotor current.

with $xk := x(kT), yk := y(kT), uk := u(kT), h(x) = i_{rd}$ and $F^T(xk, uk) = eT(Lf + ukLg)x_K$. In this case we compute a digital control law

$$u_d = u(kT) + Tw_{1k} \tag{53}$$

which solve the problem.

5.3 Simulation

The wind speed and the DFIG are shown in **Figure 6**. The estimation of the wind speed was generated based upon the nonlinear mapping of the measured output power of the generator while taking into account the loss of power in the wind turbine. The quadratic rotor current that shall be set as the reference signal

Figure 6.
Wind speed estimation.

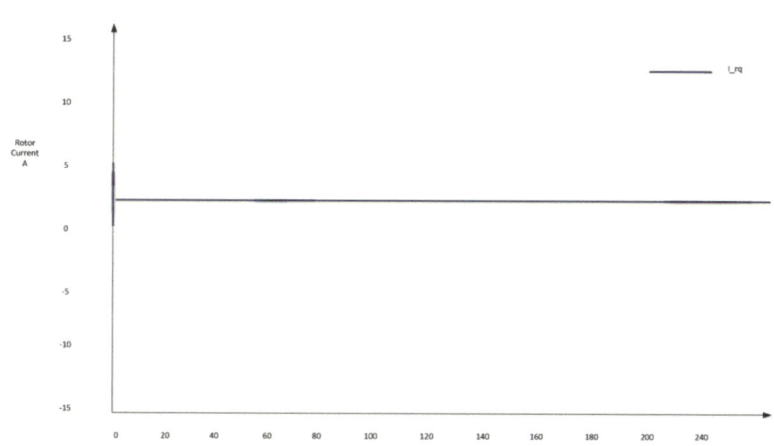

Figure 7.
DFIG rotor current.

so that a better performance is provided are shown in **Figure 7**. A feedback that is based on the proposed technique is applied and this will yield to an output signal that will follows the rotor signal (see **Figure 5**). **Figure 8** depicts the emulated and the sampled rotor speed after applying the feedback. Maximum Power Point Tracking technique was used to set the best conditions in order to arrive to maximum efficiency. It can be shown that sampled-data design provided better results such as the variation is smoother and the transient time is less than

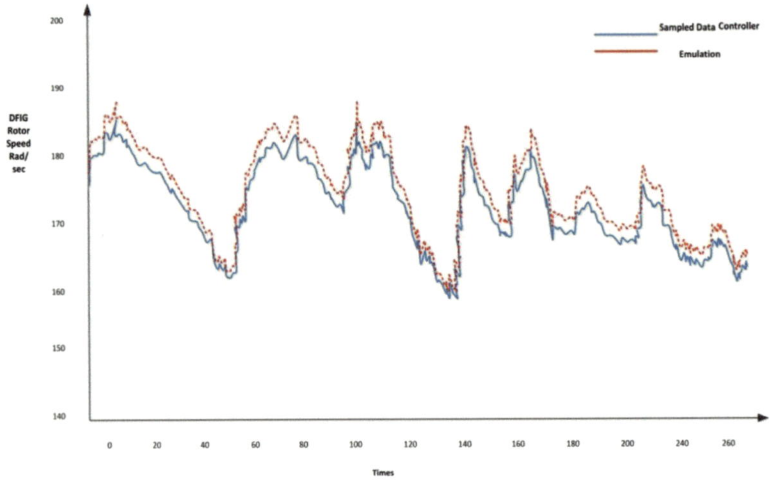

Figure 8.
DFIG rotor speed for MPPT.

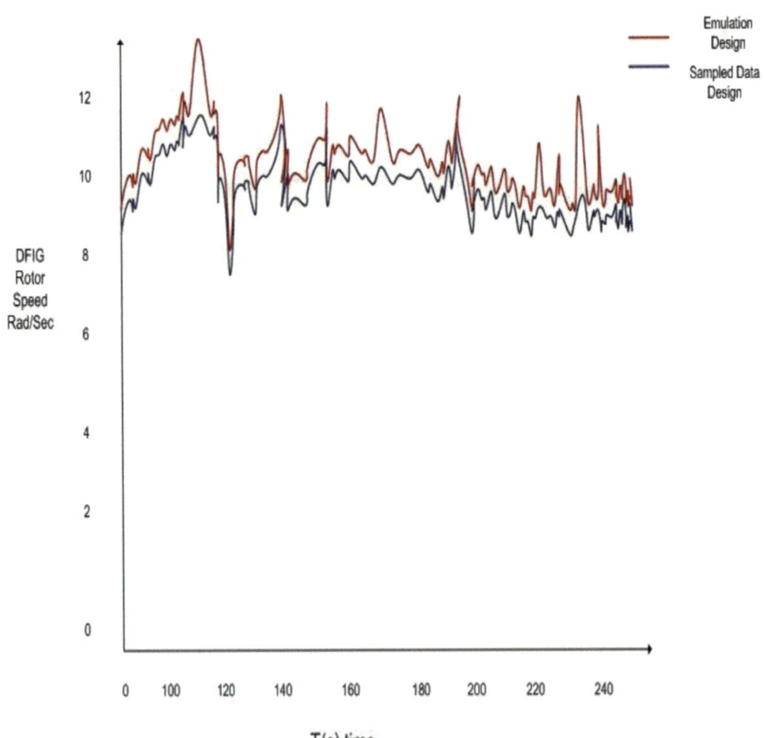

Figure 9.
Tip speed ratio.

the emulated one. The **Tip Speed Ratio** TSR is illustrated in **Figure 9** for both the emulated and sampled base. The results will show that the TSR has been reduced by more than design which indicates that the size of power converters is reduced. Then, the power converters can be downsized without reducing the output power.

6. Conclusion

This paper aims to investigate the different modern control strategies in the power system application. In particular, the study will focus on the effect of nonlinear control techniques and SampledData Model when it is applied to a Doubly Fed Induction Generator DFIG. The mechanical model was first recalled and then the nonlinear model and control techniques were discussed. In the nonlinear, the asymptotic output tracking technique was chosen where feedback is designed to ensure that the system will converge to a specific target or reference. In this case, through controlling the direct and quadratic frame we can control the active and reactive power which was proven by the results. In the second half of the chapter we choose to investigate the digital control techniques where a comparison between the emulation design and the sampled data techniques are carried out The MATLAB program was to choose to simulate and test the control strategies. It can be noted from the results that as time increased the emulation design fail to preserve the same behavior as in the continuous-time and an oscillation takes place, unlike the sampled data design. Finally, it can be concluded that applying the sampled data model over the nonlinear system provides powerful results than the classical solution. Further investigation will be carried out regarding practical cases.

Acknowledgements

All thanks to professor Salvator Monaco, my mentor.

Author details

Marwa Hassan
Sapienza University of Rome, Rome, Italy

*Address all correspondence to: eng.marwaabdelhamied@gmail.com

IntechOpen

© 2020 The Author(s). Licensee IntechOpen. Distributed under the terms of the Creative Commons Attribution - NonCommercial 4.0 License (https://creativecommons.org/licenses/by-nc/4.0/), which permits use, distribution and reproduction for non-commercial purposes, provided the original is properly cited. CC BY-NC

References

[1] Intergovernmental Panel on Climate Change (IPCC). In: Edenhofer O, Pichs-Madruga R, Sokona Y, Seyboth K, Matschoss P, Kadner S, et al., editors. IPCC Special Report on Renewable Energy Sources and Climate Change Mitigation. Prepared by Working Group III of the Intergovernmental Panel on Climate Change. Cambridge University Press, Cambridge, United Kingdom/ New York, NY, USA; 2011. p. 1075. (Chapter 9)

[2] Turner JA. A realizable renewable energy future. Science. 1999;**285**(5428): 687-689

[3] Boyle G, editor. Renewable Energy. United Kingdom: Oxford University Press; 2004. p. 456. ISBN-10: 0199261784. ISBN-13: 9780199261789

[4] Wang F, Liu D, Zeng L. Modeling and simulation of optimal wind turbine configurations in wind farms. In: 2009 World Non-Grid-Connected Wind Power and Energy Conference. IEEE; 2009

[5] Wang F, Liu D, Zeng L. Study on computational grids in placement of wind turbines using genetic algorithm. In: 2009 World Non-Grid-Connected Wind Power and Energy Conference. IEEE; 2009

[6] Khare V, Nema S, Baredar P. Reliability analysis of hybrid renewable energy system by fault tree analysis. Energy & Environment. 2019;**30**(3): 542-555

[7] Iniyan S et al. Energy models for renewable energy utilization and to replace fossil fuels. Methodology. 2020; **67**:28-37

[8] Bauwens T, Devine-Wright P. Positive energies: An empirical study of community energy participation and attitudes to renewable energy. Energy Policy. 2018;**118**:612-625

[9] Niraula M et al. Variable stator frequency control of stand-alone DFIG with diode rectified output. In: 2018 5th International Symposium on Environment-Friendly Energies and Applications (EFEA). IEEE; 2018

[10] Kashkooli MRA, Madani SM, Lipo TA. Improved direct torque control for a DFIG under symmetrical voltage dip with transient flux damping. IEEE Transactions on Industrial Electronics. 2019;**47**:777-780

[11] Errouissi R et al. Offset-free direct power control of DFIG under continuous-time model predictive control. IEEE Transactions on Power Electronics. 2016;**32**(3):2265-2277

[12] Bektache A, Boukhezzar B. Nonlinear predictive control of a DFIG-based wind turbine for power capture optimization. International Journal of Electrical Power & Energy Systems. 2018;**101**:92-102

[13] Rashid G, Ali MH. Nonlinear control-based modified BFCL for LVRT capacity enhancement of DFIG-based wind farm. IEEE Transactions on Energy Conversion. 2016;**32**(1):284-295

[14] Li P et al. Nonlinear controller based on state feedback linearization for series-compensated DFIG-based wind power plants to mitigate sub-synchronous control interaction. International Transactions on Electrical Energy Systems. 2019;**29**:1

[15] Rashid G, Ali MH. Nonlinear control-based modified BFCL for LVRT capacity enhancement of DFIG-based wind farm. IEEE Transactions on Energy Conversion. 2016;**32**(1):284-295

[16] Yang B et al. Perturbation estimation based robust state feedback control for grid connected DFIG wind energy conversion system. International Journal of Hydrogen Energy. 2017; **42**(33):20994-21005

[17] Gao Y, Ai Q. Maximum power tracking control for parallel-operated DFIG based on fuzzy-PID controller. Journal of Electrical Engineering and Technology. 2017;**12**(6):2268-2277

[18] Ponce P, Ponce H, Molina A. Doubly fed induction generator (DFIG) wind turbine controlled by artificial organic networks. Soft Computing. 2018;**22**(9): 2867-2879

[19] Yogarathinam A, Chaudhuri NR. Wide-area damping control using multiple DFIG-based wind farms under stochastic data packet dropouts. IEEE Transactions on Smart Grid. 2016;**9**(4): 3383-3393

[20] Zhou Y et al. A dynamic weighted aggregation equivalent modeling approach for the DFIG wind farm considering the weibull distribution for fault analysis. IEEE Transactions on Industry Applications. 2019;**22**:1-7

[21] Zhou Y et al. A dynamic weighted aggregation equivalent modeling approach for the DFIG wind farm considering the Weibull distribution. In: 2019 IEEE/IAS 55th Industrial and Commercial Power Systems Technical Conference (ICPS). IEEE; 2019

[22] Nallappan G, Joo Y-H. Robust sampled-data fuzzy control for nonlinear systems and its applications: free-weight matrix method. IEEE Transactions on Fuzzy Systems. 2019; **27**:2130-2139